Knowing Yellowstone

Science in America's First National Park

Jerry Johnson, Editor

Cover Photo: A Storm Ripping Through the Valley, Trey Ratcliff
DESIGN/LAYOUT/PRODUCTION – Monica Chodkiewicz

Published by Taylor Trade Publishing
An imprint of The Rowman & Littlefield Publishing Group, Inc.
4501 Forbes Boulevard, Suite 200, Lanham, Maryland 20706
http://www.rlpgtrade.com

Estover Road, Plymouth PL6 7PY, United Kingdom

Distributed by National Book Network

British Library Cataloguing in Publication Information Available

Library of Congress Cataloging-in-Publication Data Available

ISBN 978-1-58979-522-8

The paper used in this publication meets the minimum requirements of American National Standard for Information Sciences—Permanence of Paper for Printed Library Materials, ANSI/NISO Z39.48-1992.

Printed in Canada

Knowing Yellowstone

Science in America's First National Park

Jerry Johnson, Editor

Acknowledgments:

Many people are involved in the evolution and eventual publication of a book. The authors of the chapters that follow willingly took time from busy field seasons and research programs to deliver their highly technical work in a format accessible to those who are not immersed in the world of doing science. I thank them for their good work.

Several individuals supported this project from the start. John Peters is the director of the Thermal Biology Institute at Montana State University and deserves special thanks for his continued institutional and financial support. The Institute conducts and promotes research and education focused on the biology and interrelated physical and chemical processes of geothermal environments in the Greater Yellowstone Ecosystem. In addition to doing important science, the center has an outstanding record of educational outreach from K-12 and beyond. John's support for this project began with a TBI "University of the Yellowstone" award and continued through to publication.

John Varley is the former director of the Yellowstone Center for Resources at Yellowstone National Park as well as the former director of the MSU Big Sky Institute; he has been immersed in Yellowstone science for over four decades. John not only provided institutional and financial support through BSI, he was a my sounding board for ideas about the book, introduced me to some of the authors, and was always available to teach me something about the history and politics of the Park. He is an understated and gracious guy and I consider him a friend.

Trey Ratcliff generously made his collection of spectacular photographs of Yellowstone available. The cover photo is his as are several photos that appear at the beginning of chapters. Trey sees the natural world differently from most of us. His eye combined with his unique artistic expression make him one of my favorite amateur photographers. Check out his website at: stuckincustoms.com and his new book "A World in HDR". Two other photographers deserve special mention - thanks to Ken McElroy and Steve Hinch for the generous donations of their photographs.

A sincere thank-you goes out to Gerry and Wright Ohrstrom.

The two Monicas – Monica Chodkiewicz, responsible for the beautiful layout of the book, and Monica Brelsford, helped me through numerous parts of the project, were a joy to work with. I benefited from their experience producing education materials for TBI. My wife Barbara, Heather Rauser at TBI, and Professor David Parker, a colleague of mine at MSU, read the manuscript and provided me with many constructive suggestions. I value their contributions and thank them. My editor at Taylor-Trade, Rick Rinehart, brought many years of interest and experience to a book on Yellowstone. Thanks.

My parents immersed us kids into the world of nature and science when we could walk. This book is dedicated to them.

Finally, thanks to all the researchers – professional and amateur alike, who have spent many decades exploring the nuances of the world's first national park. They make the Yellowstone a richer and more stimulating place for all of us.

MONTANA
STATE UNIVERSITY
Thermal Biology
Institute

MONTANA
STATE UNIVERSITY
BIG SKY
INSTITUTE

Foreword

John W. Peters & John D. Varley

Yellowstone National Park represents many things to many people. For a select and lucky few, the Park is a place to do important and complex science. The government-sponsored and comprehensive Hayden Expedition of 1871 began a tradition of research in Yellowstone and within the first decade of the Park's existence, mechanisms were already being put in place for the management and regulation of all human endeavor in Yellowstone, including scientific work. In the early days few people were allowed direct access to park resources for scientific research -most likely because there were few scientists--but today, NPS staff issue and monitor over 200 research permits each year which makes Yellowstone one of the most studied parcels of ground in the world.

Except for brief periods when a few scientists took advantage of their privileges, the commitment of the National Park Service to research in Yellowstone has been unwavering. Even in the summer of 1988, as wildfires were still burning, scientists began designing research protocols and collecting data on burn patterns and fire behavior, impacts on park wildlife, and forest regeneration and many of those studies continue today.

Today's challenge for the park service and for researchers is not much different than it has been throughout Yellowstone's history: how to balance access to sites of scientific interest and structuring studies to be informative while avoiding ecological impact. Both sides also increasingly recognize the importance of passing on what is learned to enhance the public's understanding and experience. Park managers know that good science can enrich a visit to Yellowstone and most visitors have a strong fascination with what science can tell them about the Park. And they should. Many of the "hot-button" issues in wildlife management, bioprospecting and intellectual property rights, biotechnology, renewable energy, and global climate change are rooted in the Park and fuels lively public discourse.

Montana State University has factored strongly in the makeup of Yellowstone-centric research across all disciplines of science. The close proximity of the Park to MSU's home in Bozeman attracts world-class investigators to join our faculty and the renowned centers of excellence that have been established. The Thermal Biology Institute and Big Sky Institute are two such centers and as directors, we are proud of the role our affiliations play in Yellowstone science and policy. A recent study conducted by MSU ecology professor Dave Roberts showed that MSU received more than five times as many competitively awarded grants and at least three times as many publications on Yellowstone as its nearest competitor. The enthusiasm of the MSU faculty for Yellowstone has been infectious making it even more exciting to be involved in catalyzing the publication of "Knowing Yellowstone". The work embraces the true breadth of contemporary science in the world's first national park. We hope it is a resource that will be cherished by those interested in Yellowstone for many years to come.

Contents

v/ **ACKNOWLEDGEMENTS**

vii/ **FOREWORD** / *John W. Peters, John D. Varley*

xi/ **INTRODUCTION** / *Jerry Johnson*

1/ **CHAPTER 1** Thinking Big About the Greater Yellowstone / *Andy Hansen*

17/ **CHAPTER 2** Mapping the Last Frontier in Yellowstone National Park: Yellowstone Lake / *Lisa A. Morgan, W.C. Pat Shanks*

33/ **CHAPTER 3** Using Yellowstone's Past to Understand the Future / *Cathy Whitlock*

49/ **CHAPTER 4** Understanding Grizzlies: Science of the Interagency Grizzly Bear Study Team / *Charles C. Schwartz, Mark A. Haroldson, Kerry A. Gunther*

65/ **CHAPTER 5** Interactions Between Wolves and Elk in the Yellowstone Ecosystem / *Scott Creel*

81/ **CHAPTER 6** Brucellosis in Cattle, Bison, and Elk: Management Conflicts in a Society with Diverse Values / *Paul C. Cross, Mike R. Ebinger, Victoria Patrek, Rick Wallen*

95/ **CHAPTER 7** Fisheries Science and Management in the Greater Yellowstone Ecosystem: Ensuring Good Fishing by Preserving Healthy Ecosystems / *Alexander V. Zale*

113/ **CHAPTER 8** If You Can't Measure It, You Can't Manage It: An Ecological Approach to Weed Management / *Bruce D. Maxwell, Lisa J. Rew*

127/ **CHAPTER 9** Yellowstone Extremeophiles: The Life of Heat-Loving Microbes / *Mark Young, Jennifer Fulton*

141/ **CHAPTER 10** The Science of Storytelling: Policy Marketing and Wicked Problems in the Greater Yellowstone Ecosystem / *Elizabeth A. Shanahan, Mark K. McBeth*

The national park is the best idea America ever had.

—James Bryce, Britain's Ambassador to the U.S. 1912

Introduction

Modern science writers take us into the world of particle physics through experiments in the Large Hadron Reactor, to the depths of the oceans in remote-controlled submarines, and into the workings of the brain via functional MRIs. Good science writing (and television) teaches as it entertains. The best stories captivate us with a good plot line that just happens to be science based – finding the Titanic with high technology, loss of the rainforest and the impact on the global ecosystem, mass extinctions and wayward asteroids. What is frequently missing though is the story of how the science behind the story was done – how the story comes to be. This book tries to fill in the gap between the research question and the research findings. The setting is one of the last large ecological refuges in the world – Yellowstone National Park and the surrounding lands. The intent is to understand the work (and fun) of doing science.

Of all the great ecosystems of the world, the lands in and around Yellowstone National Park – the Greater Yellowstone Ecosystem (GYE) must rank very high in terms of the complexity of interactions between earth's natural processes, the web of life, and human actions. As the world's first national park, Yellowstone and the surrounding land attract global attention from policy makers, environmentalists, and in particular the scientific community who rightly see it as a field laboratory of boundless potential. In some instances, the research is dangerous and spectacular – trapping and tagging grizzly bears is a serious undertaking; a misstep near a hot springs can mean disaster. For others, scientific discovery takes place with the aid of high-tech remote sensors that image parts of the park unvisited by tourists. Computers and digital models turn electronic signals into a map or a simulation model that can be educational or used for management purposes. All the researchers in these chapters share a passion for a part of the world unique in its geography and natural history. All of us recreate in and around the park exploring the backcountry.

Old Faithful geyser is the undisputed personification of Yellowstone. Indeed, without the spectacular geothermal features so prominent in the Park's iconography, the region certainly would not have been considered distinct from millions of other acres in the American West. But the region is so much more. More than 23,000 square miles (60,000 km^2) demarcate the area known as the Greater Yellowstone Ecosystem (GYE). The concept of the region as an intact ecosystem was originally adopted as a formal management concept as early as 1971 by the National Park Service and was defined as the range of the Yellowstone grizzly bear by John Craighead in 1984. The GYE is consists of two National Parks (Yellowstone and Grand Teton), seven National Forests, and numerous other federal and state jurisdictions totaling 28 political jurisdictions

▶ Yellowstone National Park is located at the heart of the region known as the Greater Yellowstone Ecosystem (GYE). At 28,000 square miles, the GYE is one of the largest intact temperate-zone ecosystems on Earth. The multiple public and private jurisdictions make comprehensive policy making for maintaining ecosystem functions particularly difficult.

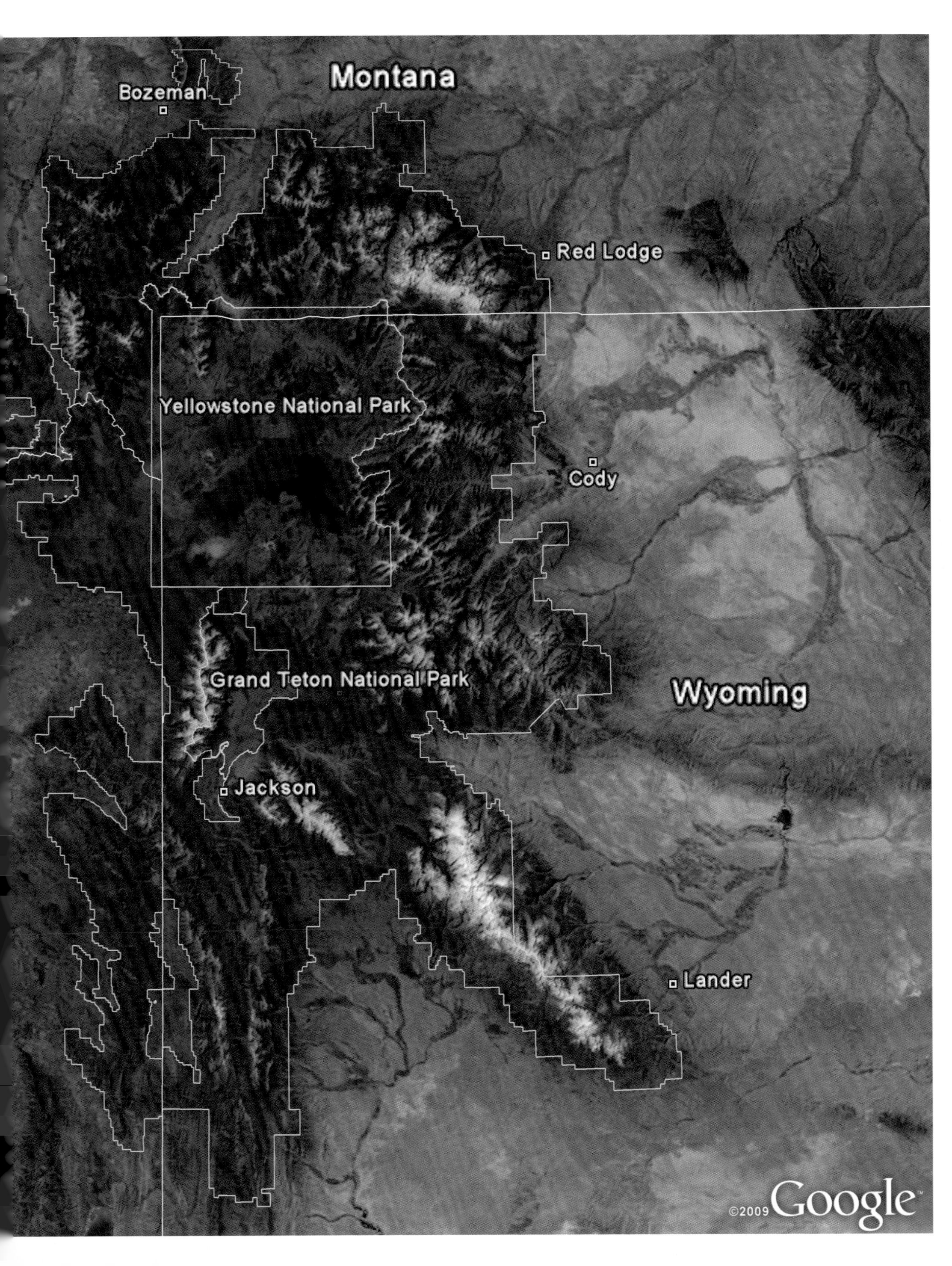
Montana
Bozeman
Red Lodge
Yellowstone National Park
Cody
Grand Teton National Park
Wyoming
Jackson
Lander
©2009 Google

in all. The management paradigms for the different jurisdictions is part of what makes the area interesting because of their different science and research needs.

The public lands in the region act as ecological and recreational refuges for almost a half a million residents and 3.5 million yearly visitors to the area. These lands are home to key predator species (grizzly and black bear, gray wolf, mountain lion), prey species (elk, deer, moose), and a host of birds, smaller mammals as well as a mosaic of vegetation including coniferous forests, arid shrub lands, and grasslands. Several major river systems originate in the region and elevation gradients range from lowland river valleys to the Grand Teton at over 13,000 feet (>3900 m) in elevation. About 70% of the land base is publicly owned with the remaining private land in agricultural production or urban and exurban development.

In part because of the legacy of Yellowstone as the world's first national park, and in part because of natural quality of the Yellowstone area, there is a rich and varied history of doing science in the region. In 1863 Walter DeLacy explored the Shoshone and Firehole geyser basins and in 1871, the geologist Ferdinand Hayden led an expedition into the heart of Yellowstone where they explored, mapped, photographed and described the natural history of the region.

The science of natural history of Yellowstone was introduced to most Americans by the National Geographic films of Frank and John Craighead in the early 1960s. Many of us remember the early films as part adventure and part nature show. I will never forget the scene of a partially drugged male grizzly charging the station wagon that served as the Craighead research vehicle. The two biologists, though their films and writing, helped us to see natural and protected areas as more than curiosity attractions for tourists; they taught the nation that concepts like biodiversity and wild habitat mattered to our national culture. They helped shape management paradigms based on preservation of large expanses of land rather than individual species. Today, these same values reside in virtually every public lands agency. Mostly, they helped create a culture of science in the Park that continues today.

Science in Yellowstone

The modern issues facing areas like Greater Yellowstone are problems of immense complexity. Contemporary management of our public lands requires research in the study of both natural and human ecology. Addressing those issues requires knowledge both deep and wide. The natural world and human world are inexorably linked and real understanding of policy solutions will not take place without recognition of those linkages. Good science is a requirement for good decision-making.

One problem for public managers is that they frequently lack broad training across disciplines and the scientific methods each employs. The same is true for those with a passion for Yellowstone – people don't always understand what they are seeing (stories of those misperceptions are legend among park researchers). Our worldview is often incomplete because we don't understand the scientific perspective of those who would inform us. Often, we simply cannot grasp the scope of the issue. An example makes my point.

Like many rural areas in the interior West, the social and economic landscape of the Greater Yellowstone Ecosystem is undergoing rapid change. The traditional resource extraction economy is being augmented and in many cases overwhelmed by a new economy based on tourism, personal services, and retirement. Meanwhile, the regional agricultural economy is stagnant.

Tourism visitation acts as a catalyst for people to relocate to rural communities; new business opportunities emerge with more people. Many of the new residents are urban refugees fleeing the crime and social problems of urban centers. They, in turn, bring new sources of income, cultural values, and inevitably impact the environment as they live, work, and play in the rural countryside.

One impact is that new arrivals have a tendency to build homes near the forest edge or in the large meadows ubiquitous in the region - places where multiple native species - birds, deer, elk, and bears, would also like to live. The setting for many of these homes is spectacular but they often have multiple impacts to the region.

The animals are displaced to perhaps less desirable and thereby less productive habitats at higher elevations or deeper into the forest, sometimes into the subdivisions themselves. Once subdivisions occupy land formerly available to herds for grazing, the result may be fewer native species, an increase in the incidence of non-natives, and conflicts between humans and predators. Today, it is not unusual for deer, elk, and even moose to live in rural subdivisions and for large predators like mountain lions to follow them there. Animals are hit on roadways; predators like bears use outside barbeques or bird feeders as food sources. Area old-timers and newcomers clash over values and property rights.

Homes on the edge of the region's lodgepole pine forests impact wildfire policy and the resources devoted to fighting it. Agencies are under intense pressure to extinguish fires before expensive homes in and near the forest are destroyed. However, the lodgepole pine common to the region is a serotinous species; their cones burst open and disperse seeds when heated by fire. Without wildfire, the forests are not rejuvenated with seedlings nor are they are not cleansed of insects like the mountain pine beetle. Without fire, deadfall trees create unusually large fuel loads. The inevitable forest fire is often large and beyond the management capacity of public lands agencies to control. The result is that political forces prevent fire from carrying out its ecological function. As ecological and political goals clash, an administrative catastrophe predictably follows as county commissioners and forest managers disagree about the proper role of wildfire near rural communities.

The inevitable lessons of ecology, politics, and social change resulted in a literal and figurative firestorm in Yellowstone over twenty years ago. The summer of 1988 was the driest in the Park's recorded history. In mid July, when several large fires began to burn in and around the Park, no one thought they would burn until late September and scorch almost 1.2 million acres (485,000 hectares) across the ecosystem. On the worst single day of the fires, August 20, tremendous winds swept fire across more than 150,000 acres (60,702 hectares).

Many were surprised at the scope and intensity of the event and some complained that the "let burn" policy of the Park was to blame. Beginning in 1972, park managers instituted a natural fire management regime where lightning-caused backcountry fires would be allowed to burn without major interventions if they met certain criteria. Most (86%) such fires burned themselves out; this complex of fires didn't. It wasn't until snow fell that fall that the fires were extinguished.

Park management came under tremendous pressure as the world's media converged on West Yellowstone to cover "the destruction of the world's first national park". The reaction of some Montana and Wyoming politicians was swift and predictable. Congressional hearings were held and resignations of the National Park Service director William Penn Mott and the superintendent of Yellowstone, Bob Barbee. In the end, Yellowstone officials modified their fire policy to continue the "let burn" management with stricter guidelines about when to intervene in nature caused fires.

After the fires of 1988, a large research effort was launched to understand the role of fire in the ecosystem. A basic finding was that the fires burned across the landscape in a mosaic of burned and unburned patches. In fact, the burn pattern added to the diversity of the landscape that is evident today. In some areas of the park there is a large variance of the age of lodgepole pine suggesting that some stands were subjected to low intensity fire while other, mature stands, burned and later reestablished themselves as seedlings. Meadows were regenerated as grasses and perennials sprouted the following spring. New aspen groves established themselves in the thicket of deadfall of partially burned trees. Any objective observation of the park today would find that the landscape is hardly "devastated" and that the new growth has been good for plants and animals alike.

After the initial shock of seeing large swaths of blackened trees, tourists found the fires to be an interesting feature of the park and an opportunity to learn about the ecology of the region. The Park responded with self-guided trails that interpreted fire ecology. Many locals do not hold the same view. Even today many continue to argue for strict control of all wildfire both in and out of the Park in order to protect forest resources and private property. Politics rather than science still govern many fire management decisions in the region.

The basics of the intricate relationships described above are relatively well understood by the scientific community at universities and within agencies that regularly manage wildlife. Rarely do well-intentioned decision makers or members of the public understand the issue in scientific terms or the limitations that naturally result from the study of complex systems in general. Maybe better understanding of the science of natural landscapes will lead to policies that mitigate our effects and result in more logical public policy.

Yellowstone was and is still an ideal place to conduct research on bears, tourists, or elk. For other disciplines, chemistry for example, Yellowstone's unusual thermal features make it a unique research setting unlike any in the world. For applied sciences, the surrounding national forests are places to run natural and purpose-designed experiments on weed control or recreation impacts. Hopefully, the following chapters present a fair summary of how scientists do their work and the strengths as well as limitations of science and scientific knowledge. Let's see what some of those scientists have been up to.

Outline of the Book

The following chapters present an array of methods across many temporal and spatial scales. They take us from the depths of Yellowstone Lake, back in time to a prehistoric Yellowstone landscape, to the high-tech analysis of grizzly bear biology. They tell us how we might better understand the myriad of seemingly intractable issues facing the GYE. Although no book on the Greater Yellowstone can present all the research taking place in the region (there are currently over 200 active research permits in Yellowstone every year) these chapters represent the breadth of topics and techniques of some of the best science in the Park.

The language of science is sometime difficult to understand and I have tried to minimize the specialized jargon of various disciplines. To the extent it was possible, I defined most terms in the context of the chapter. In many instances sidebars in some chapters provide more in-depth coverage of a technology or terminology. The captions of photographs and technical figures are intended to be comprehensive and add to the understanding of the chapter. You will also notice that all measures of distance, volume, and elevation in each chapter are presented in the metric system - the language of science. Finally, each discipline and author has a "voice" that I tried to preserve as I edited individual chapters.

Several themes run through the chapters. One is climate change. The initial design for the book did not include a chapter explicitly on climate change; I felt the book should focus on specific policy issues local managers grapple with on a daily basis - predators, social and economic issues, invasive species. However, as authors began sending in their chapters it became apparent that climate change was potentially a major vector for change in the Greater Yellowstone. It seems clear that global climate change will inevitably have some impact on the Greater Yellowstone Ecosystem and even though there may not be much we can do about it in the short run, I think the chapters treat the numerous potential impacts of climate change in a way that outlines how the region might be affected.

Another theme is the close linkage between fieldwork and technology. A popular (mis)conception of science is that of the lone researcher in white lab coat hunched over a microscope. The reality in Yellowstone is much different. Olaus and Adolf Murie's ecological work in the 1930's in Yellowstone and Jackson Hole represents the traditional way of doing field ecology in complex systems. They lived in their study sites, sometimes for many months, observing the processes of the natural world. Few contemporary scientists have the professional luxury of dedicating such large amounts of time to their research activity. Today, field researchers, and everyone in this book is an outstanding field researcher, spend many hours and even days watching a large predator, taking samples from hot springs, or probing deep in Yellowstone Lake. In some cases they collect observations over many years, even decades. But, one advantage today's scientists have are the powerful computers and sophisticated software systems that shortcut the work of assimilating large data sets. One of the benefits is that disciplines can now more easily work together on big research questions and knowledge moves forward at an astonishing rate.

The most important technological advance for the study of large natural areas like Yellowstone depends on an array of 24 satellites originally deployed by the US Department of Defense. In the 1980s, the "Global Positioning System" was made available for civilian use. GPS allows the user, through triangulation of low power radio signals from two or more satellites, to know his location almost anywhere on earth. This capability allowed for the development of advanced Geographic Information Analysis, radio tracking of study animals, deep-water imaging, and a host of other space age technological methods. These methods do not obviate the need for fieldwork but they do add tremendous value to it.

Finally, there is a great deal of uncertainty in the study of large natural systems and there is still a lot we do not know about Greater Yellowstone. The level of complexity in large natural systems is often difficult to communicate to nonscientists and the result is frequently a muddled understanding by the media and interested public. No responsible scientist would claim to know how climate change would affect the region and its inhabitants but we know it will. Researchers are still sorting out what role wildfire plays in ecosystem health and how to integrate fire and private land management. It may be some time before we know the full ecological and social effects of maintaining large herds of ungulates and how to do so even as we manage for large predator species. A looming unknown is what the future holds for an ecosystem experiencing double-digit population growth rates in the counties that surround the Park.

The book is divided into parts based on the logic of understanding a large ecosystem. The first three chapters examine efforts to describe the terrestrial landscape – the landscape ecology of the region and the role of humans, the unseen geography and geology of Yellowstone Lake, and the landscape of time. They set the stage for the six chapters that follow – the study of the wildlife ecology of the region and how we study life forms from the Yellowstone Grizzly to the invisible life in hot springs and thermal vents. The last chapter treats the study of humans in the ecosystem as a complex matrix of political actors, value-laden individuals, and media coverage. Hopefully, the reader will explore connections between chapters and begin to understand the Greater Yellowstone as a larger system of natural and political processes.

Chapter 1

Thinking Big About the Greater Yellowstone

Chapter 1

Andy Hansen begins with the "big picture" look at the Greater Yellowstone Ecosystem and how landscape ecologists do large-scale ecological description and analysis. Landscape ecology finds its antecedents in regional geographic studies in 1930s Europe where the emphasis was on describing how humans had modified the landscape. Today, the emphasis is on understanding how the landscape interacts with all parts of the system - the climate, topography, hydrology, and human and non human agents. Taken individually, each detail of the Greater Yellowstone is an interesting story of natural history and ecology but the best way to appreciate the full complexity is by understanding how the parts work together to form an intact and functional ecosystem.

Landscapes with a large amount of variation within them present public land managers with especially difficult management issues. The volume of information needed to design management interventions that are effective require large scale thinking and fine detail science. Hanson is uniquely qualified to understand landscape scale inquiry. He has done an immense amount of work in Yellowstone but also in Alaska and Serengeti National Park in Tanzania. Hanson's approach is to select species that can be studied relatively easily and in large numbers - in this case forest vegetation and birds. The advantage to this approach is that he can sample across multiple land cover types along large elevational gradients.

The unique part of Andy's work is the melding of social data - housesites and land use, and ecological data - bird nest sites and success. He is among a handful of ecologists working in the Rocky Mountain west taking a comprehensive look at the impacts of human population growth on the rural countryside. Andy's work, along with others, suggest that some of our impacts, beyond the obvious ones of decreasing open space and crowding riparian zones, have unseen cascade effects. Our homes and agriculture attract populations of native birds and their predators. Hanson found that the predation may affect native bird populations many miles away.

The effects of homes in the countryside have impacts on wildfire policy, endangered species, and ecosystem services. These and other issues will be increasingly important as the human population of western communities continues to grow and change rural landscapes.

J. Johnson

Gateway to Yellowstone (Trey Ratcliff)

Chapter 1

Thinking Big About the Greater Yellowstone

Andy Hansen

Andy Hansen, Department of Ecology, Lewis Hall, Montana State University, Bozeman, MT 59717; Email: hansen@montana.edu

Visit Andy's landscape ecology lab at: http://www.homepage.montana.edu/~hansen/index.htm

"How does it look, Jerry?" Jerry Johnson had just side slipped out of sight down the summit snowfield checking out the route for our descent. Bruce Maxwell and I were finishing lunch on the top of Hummingbird Peak in the Lee Metcalf Wilderness on the northwest side of the Greater Yellowstone Ecosystem.

The July morning air was crystal clear. The view was dominated by the steep, snow-covered Spanish Peaks and the long runs of conifer forests spilling off their sides. We had seen elk, moose, mule deer, black bear, and possibly a grizzly bear on the drive through a large private ranch into the trailhead. Like most places in Greater Yellowstone, this place is big and wild.

To the north, we could see farms, rural homes, and subdivisions dispersed among the cottonwood trees on the Gallatin River floodplain. The fertile soils of the Gallatin Valley attracted a few hearty farmers in the late 1800s and the valley's population remained small until a wave of immigration started about forty years ago. Many of those newcomers wanted to live "out of town". We could see their rural homes and ranchettes sharply delineating the boundary of the Gallatin National Forest.

Among the tall mountains to the south stood Lone Peak. In the 1970s, Chet Huntley, the famous retired newscaster, built a ski area in this high rocky country. In recent years, the real estate market has flourished around the Big Sky Ski Resort. We could see the Rocky Mountain version of mansions perched on the steep forested slopes of the mountain.

"Bruce, most of the houses we see were not here when you were a kid in Bozeman. How do you think they influence the wildlife and ecosystems in Yellowstone National Park and the surrounding wilderness?" This was the question that many scientists were asking as more and more people are moving in around the fringes of the Yellowstone wilderness.

"Steep, damn steep!" Jerry's voice punched into our conversation. This got my attention because Jerry is so solid on skis that I never heard him say this before. Bruce jumped to ski the headwall with Jerry. I skirted left to a slope that was just steep. We then did turns down a 457 m vertical drop and then hiked some 16 km to the car.

In a place as large and wild as greater Yellowstone, it might seem silly to ask about the effects of the growing, but still small, human population. Greater Yellowstone is often referred to as the largest "intact" ecosystem in the coterminous United States. Yellowstone National Park, at 8,987 km^2 is the largest national park in the lower 48 states. As large as YNP is, however, it represents less than 10 % of the Greater Yellowstone Ecosystem, the larger surrounding ecosystem that YNP and Grand Teton National Parks are dependent upon.

Here is what "big" means. Within the GYE is Yellowstone Lake, the largest lake above 2438 m in the lower 48; the Beartooth Plateau, the largest area above 3048 m in the lower 48, and Two Oceans Pass, close to the place that is the most distant from a road (ca 53 km) in the lower 48. Greater Yellowstone is about 482 km north to south and 450 km east to west. According to MapQuest, the time to drive around the 1320 km perimeter of the GYE is more than 17 hours.

Currently, around 425,000 people live in the GYE. Because of the large size of the system, the population density is only 2.93 people/km^2, very low compared to most places in the lower 48 states. The U.S density is almost 30 people/km^2. Sixty eight percent of the ecosystem is publically owned; residents live mostly in the perimeter ring of the system, leaving the vast interior as a wilderness with more grizzly bear, wolves, and elk than humans.

The current low density of people, however, is twice the density as it was in 1970. Lewis and Clark were the first EuroAmericans in the area in 1806. Blackfeet, Crow, and other Native American tribes continued to control portions of the area until the mid 1870s, soon after YNP was established. Euro-America population growth over the next century was relatively slow due to the area's harsh climate, limited agricultural opportunities, and distance from major cities. The same factors that dissuaded population growth prior to 1970, however, now act as attractants to the region. As part of the environmental movement, many people left the cities looking for places with wilderness, scenery, outdoor recreation, and other "natural amenities". The Internet and increased wealth allowed many people and businesses to relocate to what they considered to be very desirable places. Thus, there has been a wave of rural home construction around the perimeter of the public lands of the GYE. These trends have led many scientists and conservationists to ask if and how rural homes at the edge of the wildlands might influence wildlife and ecosystem processes in Yellowstone and Grand Teton National Parks.

How would you answer this question? In a place so big and wild, how would one assess the effects of homes sprinkled along the vast perimeter of the Yellowstone wildlands? This is a question I have been working on over the past 15 years. In this chapter, I will share the variety of methods that my colleagues my students helped me develop to investigate large-scale ecological impacts in and near protected areas. Some of them are based on high technology but as we shall see, getting to know the backcountry with other scientists from an array of disciplines and mobilizing large field crews is one of the key methods.

Overview of Methods

In a large wild system like Greater Yellowstone, a first task is to quantify where wildlife and rural homes are located across the landscape. With this information, specific places can be used to study how various types and densities of rural home development influences wildlife. Finally, these effects can be summarized across Greater Yellowstone to draw conclusions for conservation. Some of the data needed for this mapping are easier to obtain than others. Satellite data can be used to map land cover and use (e.g., agriculture, cities, vegetation type) over large areas and such land cover products are now available for free download from the internet; these data are readily incorporated into GIS software (see sidebar). Rural homes are too small to be detected with the satellite sensors regularly used but homes can be seen on aerial photographs and mapped accordingly. However, hundreds of photos would be needed to cover the GYE and this method is prohibitively expensive. County governments record permits for the water wells that are typically drilled at rural home sites and attach geocodes (latitude/longitude) to the well location. They also record individual land tracts for tax assessment. By going to the 20 counties of Greater Yellowstone, one can obtain these data and map the density of rural homes. Data on wildlife typically must be collected in the field. The methods can be very time consuming and thus relatively few species are well studied. Moreover, wildlife abundance, birth, and death vary across landscapes, so studies must be done in many landscape settings. The effects of rural homes on wildlife vary with home type, home density, and location. Putting all these data layers and interactions together to answer what might seem a simple question is challenging.

Studying Spatial Patterns of Biodiversity

A place like Greater Yellowstone supports many, many species of plants and animals. Which species should be included in studies of rural home effects? We use the term 'biodiversity' to refer to the full range of life in an area. This term is typically defined as, the variety of all forms of life, from genes to species, through to the broad scale of ecosystems. Most people think of grizzly bear, elk, or bison as emblematic of Yellowstone. As interesting as these animals are, every species tends to use the environment is different ways and at different scales. Thus, the response of one species to a given type of land use may be very different than that of other species. Consequently, MSU colleague Jay Rotella and I selected for our initial studies taxonomic groups for which several species could be quantified with one set of relatively cost effective methods. These groups were birds, trees, and shrubs.

Trees and shrubs are stationary and thus are easy to

Geographic Information Systems

Of all the sophisticated analytical techniques employed by the researchers in this book, none compare to the impact of Geographic Information Science (GIS). Every natural and social science benefits from the ability to accurately identify data with a location on earth.

Contemporary digital mapping and analysis is made possible by the deployment of the Global Positioning System (GPS), a worldwide system of 24 satellites and their ground stations. These "man-made stars" use a radio signal and triangulation to calculate positions accurate to a meter in most cases and, in others within centimeters. Once we use GPS to know where we are, we can use GIS software to reference data to the known point; those data can be used for analysis between and among points.

GIS technicians generally utilize three methods to analyze data: mapping, relational data analysis, and modeling. All three methods can be combined in order to understand immensely complex systems like Yellowstone. Here are some examples.

Maps can be constructed from data if it is anchored with a spatial reference. Ecologists frequently use radio collars on animals and use GPS to track them across the landscape. GIS mapping allows them to view animal habits in the context of land cover and elevation as well as the proximity to food or human infrastructure (i.e. roads). By viewing animal habits geographically ecologists gain new insights into behavior.

Many ecological processes (land cover, fires, habitat) are related to geography. Discovering patterns in relationships is known as relational data analysis. Soils, precipitation, and land cover are clearly related and produce known ecological landscapes; the resulting landscape mosaic will help determine the suite of animals attracted to the area and thus, we can build a multilayered representation of an ecosystem. Changes to the mosaic due to, for example, wildfire, may give researchers insights into how ecosystem processes will change. Other layers can be added that represents human caused changes to the landscape such as roads, development, or human activities. In this example, GIS technology could be used to discover how roads or development might impact wildfire behavior and how that might change habitat for particular species.

Models can be constructed from the data to, in effect, create new data. If the data stored in the computer is robust enough, researchers can use the software to imagine a different reality - one where the rate of change is slowed or accelerated, for example. Modelers can then make forecasts about plant or animal populations or how various management scenarios might change the future of the resource.

---

tally with forest surveys. Many species of birds set up small territories during the breeding season and sing to defend the territories from individuals. Thus, their abundances can be easily surveyed by visiting a location and recording all species seen or heard for a fixed period of time. Through such methods, the abundances of many species can be quantified. Because each species makes a living in a slightly different way, such studies can reveal the variety of ways that land use may influence species and communities.

While abundance of bird, tree, and shrub species are rather easily sampled, estimating population growth or decline requires knowledge of birth rates, death rates, and movement. These are difficult to near impossible to measure for multiple species. We focused on birth rates for several bird species and made assumptions about death rates and movements based on studies in other locations. Measuring reproductive rates on birds requires finding their nests and recording the number of eggs or hatchlings. One returns to each nest every few days until the young die or fledge from the nest. Finding these nests in dense forests takes a special knack and lots of patience. Returning to all nests every few days must be done carefully so as not to attract predators.

Because monitoring wildlife populations is so challenging, the traditional approach is to collect data in several locations and assume that these samples were representative of all places across the landscape. However, ecologists have increasingly learned that populations sometimes vary across landscapes based on habitat quality, food availability, climate, predators, and other factors. This is especially true for places like Greater Yellowstone that have pronounced gradients in elevation, climate, soils, and other factors. Given that it is not feasible to measure every place in the ecosystem, how can we characterize these "spatially explicit" population dynamics?

We attempt to deal with this complexity by quantifying "patterns of association" between wildlife populations and environmental features that influence population dynamics such as proximity to water, road density, rainfall, and slope aspect and angle. Some of these environmental features have been mapped so we use statistical approaches to predict species abundance or reproduction for each location across the landscape based on the value of known environmental feature. In this way, what is learned at field sample sites can be "painted" over the full area of GYE based on controlling factors such as climate or plant productivity.

Birds Across Greater Yellowstone

Our first studies in GYE focused on the northwest portion of the ecosystem including the Gallatin, Madison, and Henry's Fork watersheds. This area was chosen because it included the major ecological zones and land use patterns typical of GYE and was of a manageable size for field sampling. Stretching some 200 km north to south, the area was still offered substantial challenges. We hired a crew of 15 technicians and they were housed in three locations across the study area: Bozeman, West Yellowstone, and Island Park.

Many bird species select habitat for breeding based on local vegetation conditions and on availability of foods such as insects, seeds, and fruits. In the Northern Rockies, habitat types, soils and ecosystem productivity are broadly related to elevation. Thus, we choose to stratify sampling for birds among the major habitat types in the area and among three elevation zones.

This resulted in field plots being distributed among riparian forests on fertile soils in valley bottoms, aspen and Douglas-fir forests on moderately productive midslopes, and lodgepole pine forests on infertile soils at higher elevation. Douglas-fir and lodgepole pine forests are subjected to disturbances including logging and

▲ PHOTO 1.1 Once bird nests are located, they are revisited every two to four days to monitor the fates of the eggs and chicks. We try to vary the path to the nest as well as the time so we do not lead predators to the nest location. (Landscape Biodiversity Lab, MSU)

▲ PHOTO 1.2 If a nest is far up in a tree we will employ a mirror or small video camera on the end of an extendable pole. This is the nest of a Yellow Warbler in the Gallatin National Forest. (Landscape Biodiversity Lab, MSU)

fire. Thus we placed samples in forest stands recently disturbed by fire and logging, mid-successional pole stands, and mature and old-growth stands. This complex sampling strategy eventually yielded 99 sample sites.

The field crews were equipped with four wheel drive pick-up trucks, mountain bikes, kayaks, and good hiking boots. Even so, we restricted sample sites to those that were within 3 km of a road to maximize the number of sites the crews could sample.

For the first three years of the study, the crews focused on sampling bird abundances and vegetation. Each morning during the bird breeding season (June-August), the crews would rise at 3 am, drive, bike, and hike to the sample areas by sunrise (ca 5 am) and begin sampling birds. Censusing the birds was the fun part of the study. As morning light begins to brighten the darkness, an explosion of bird song begins. Crew members frantically record all birds species they see or hear for a 10 minute period. It takes special skill to immediately identify the individual songs of the 60 or more species that might be encountered and the crew had spent the month of May training their ears to the birds using digital recordings. After a point count was completed, the recorder would charge 200 m through the forest to the next point and begin another count. This would continue for three or four hours until the "dawn chorus" of bird songs concluded for the day.

After a bit of "lunch" at 9 am or so, the crews shift over to sampling the vegetation at each bird count point. Now the more tedious work would begin. Crews tallied by species and size class all trees, shrubs, and herbaceous plants in one to eight meter radius plots distributed around the bird point count locations. This could involve hundreds to thousands of stems. By days end at 2 pm or so, the crews made the long trek and drive back to the field headquarters. By rights, the crews should have fallen exhausted into their bunks. Being college undergraduates with lots of energy, however, they typically went fishing, climbing, or white-water kayaking until dark.

The last two years of the study focused on estimating bird reproduction. Each crew member had two large tracts of forest where their task was to locate bird nests and monitor the fate of young in the nests. Most of these bird species attempt to avoid nest predation by hiding their nests from the ever-present crows, ravens, magpies, and squirrels that make meals of eggs or nestlings. Finding these heavily camouflaged nests takes great skill and patience. Some days, a crew member would find one nest in a 6-8 hour search. Other days they might find 8 or 10 nests. Once the nests were located, the crews enjoyed returning to them every 2-4 days to record the number of eggs or young remaining. Often, crews had to use mirrors on extendable poles to view inside the nest from below. It was especially fun later in the season to see nestlings make their first flights as they fledged from the nest.

Working in the backcountry of Greater Yellowstone brings a wealth of potential dangers and hazards. One team awoke in their tent in the night to find that three feet of snow had fallen and that avalanches were crashing down the nearby slopes above the tent. Another crew watched a bull bison jump a five strand barbed wire fence when it charged them. Most harrowing was a grizzly bear encounter. The large bear appeared a few meters from a crew at a point count station. One new member of the team panicked and began to run away, a sure way to cause a grizzly to charge. Fortunately, the more seasoned companion grabbed and held him as they both slowly backed away from the bear. The bear followed them keeping within several meters for the terrifying one kilometer march to the truck. Fortunately, none of the crew was injured over the course of the study and a large quantity of data was collected.

After the field season was concluded, the data were entered into computer databases, subjected to quality controls and statistically analyzed. The goal was to summarize patterns of species abundances, number of bird, tree, and shrub species, and bird reproduction among the biophysical settings and habitat types sampled. A variety of statistical techniques are needed to determine how these measures of biodiversity varied with elevation, soils, habitat type, seral stage of habitat, and local "microhabitat" conditions. The resulting "statistical patterns of association" are especially important for studies of large landscapes like ours in GYE. They can be used in reverse to predict biodiversity patterns for places where birds, trees, and shrubs were not sampled. Because we had maps of elevation, habitat type, soils, and seral stages, we were able to predict based on these factors the abundances of species, numbers of species, and bird reproductive rates for all parts of the study area with a known level of confidence.
This set of methods for "painting" biodiversity across the landscape involve high technology tools including remote sensing, computer-based geographic information systems, stratified-random sampling designs, and statistical techniques. However, a key ingredient to producing good results is old-fashioned "knowing the land". Our backcountry forays on skis, mountain bikes, and kayaks are aimed at coming to know how the ecological system and the wildlife species are interrelated and patterned. Being intimate with the ecosystem allows us to select the right study areas and sampling methods to produce results with high levels of confidence.

We learned from this study that wildlife species were by no means evenly distributed over the landscape. Quite to the contrary, most species were concentrated is small areas that we call biodiversity "hotspots". They tended to be in valley bottoms with fertile soils, adequate ground water, warmer summer temperatures, high net primary productivity, and deciduous woodland habitats. These cover only about 3% of the study area. They are mostly outside of Yellowstone National Park and outside the national forest. They are primarily on private land. The explanation is that most of GYE is higher in elevation, has long harsh winters with very short growing seasons, poor volcanic soils, and lower net primary productivity. Thus, the well-developed habitats, warmer temperatures, and more abundant foods favored by many bird, shrub and tree species are scarce over most of the landscape. It is the more mesic - "moderately moist" valley bottoms, largely outside of the public lands, where these conditions occur and many wildlife species are concentrated.

Ecosystem Productivity

Ecosystems are composed of individual plants and animals and the physical factors they require such as soil, water, and nutrients. The collective interactions of organisms and environment result in emergent properties of the ecosystem. One of these is ecosystem productivity. This is a general term for the amount of energy that flows through the ecosystem. Sunlight is converted by primary producers (green plants) to organic molecules in the form of leaves, wood, fruits, seeds, etc. Primary consumers (herbivores) eat these plants and convert the energy to animal protein, which may then be consumed by secondary consumers (predators). Ultimately, dead organic matter is consumed by decomposers. The amount of new biomass fixed by plants per unit time and area is called net primary productivity (NPP). NPP is critical to the birds and mammals that live in the ecosystem because it sets the total amount of food available. Consequently, the number of species in an ecosystem and the abundance of each are often related to NPP. NPP varies across the landscape with climate, soils, and other factors. These spatial patterns of NPP influence the distribution of biodiversity across the landscape.

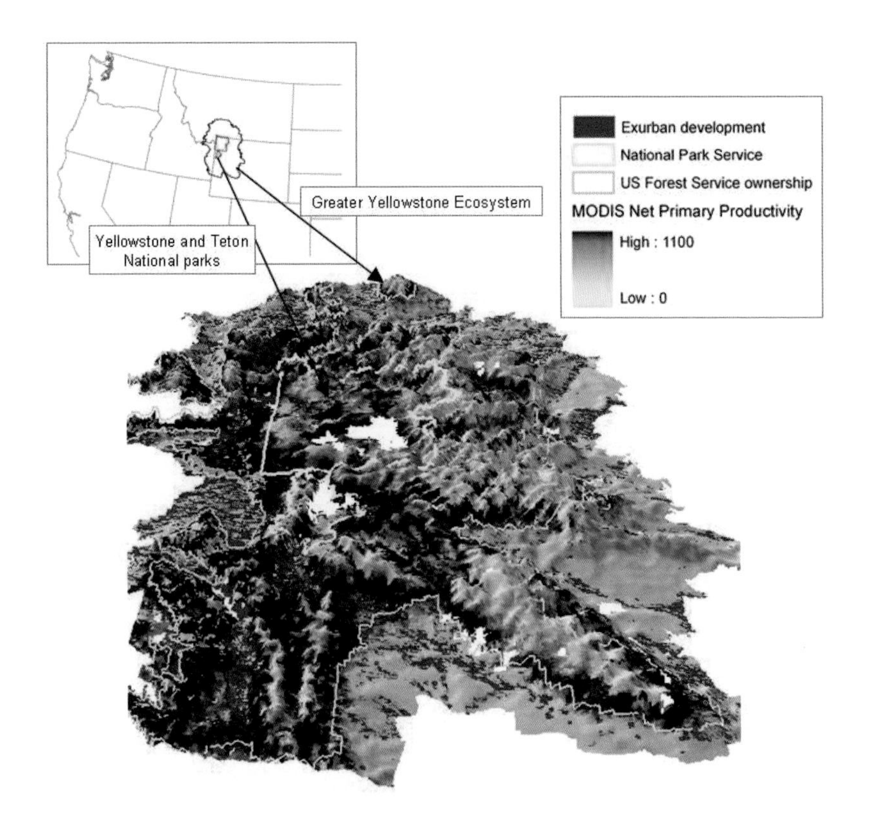

▲ FIGURE 1.1 The highest biological productivity in the Greater Yellowstone Ecosystem is found along the edges of the forest boundary and the fertile river valleys - places that also attract people. Often, in the case of national parks like Yellowstone, the least fertile ground enjoys the highest levels of protection but they are not the landscapes with the highest species richness. Expansion of land conservation strategies captures larger areas of relatively high productivity in a protected areas system and leads to greater protection of biodiversity. (Landscape Biodiversity Lab, MSU)

Spatial Patterns of Land Use

Land use describes the ways that people live in, travel across, work and play on the landscape. Data on some classes of land use can be drawn from readily available maps of cities, roads, and land allocation boundaries. Other classes of land use, such as location of agricultural fields, must be mapped from satellite imagery, aerial photographs, or other data sources such as US census data. No one had previously mapped the distribution of rural homes across the vast GYE. We had to use innovative methods to generate the first such maps of rural homes.

Andrea Parameter led the first step, which involved mapping as many land use classes as possible from Landsat satellite imagery. These classes included urban areas, irrigated crop lands, dry-land crops, and timber harvest units. We additionally mapped natural vegetation types including grasslands, shrublands, deciduous riparian forests, Douglas-fir forests, and lodgepole pine forests. The Landsat satellite sensor records the reflectivity of light from the land surface in seven wavelengths called spectral bands. A sensor on the satellite records these data at a 30 m spatial resolution; each of the land use and land cover classes reflects these seven bands in different intensities that can be represented as colors in the lab. Only one to four cloud-free images can be obtained per year. Before we can load the satellite data for analysis we frequently need to calibrate the "spectral signature" of each land cover and use class with observed data from aerial photographs. The photographs have a spatial resolution of about 1 m and the land cover and use classes can be identified by eye on the photos. We randomly select on the photos

▲ PHOTO 1.3 Many landscape features such as habitat types and land use patterns can be detected by eye on high resolution aerial photographs. The resulting data can be used to both calibrate use of satellite data for quantifying landscape features over large areas and for determining the accuracy of the satellite-derived maps. (Landscape Biodiversity Lab, MSU)

for the study area some 30-50 locations of each land cover and use class. These samples are collocated on the satellite imagery and the spectral properties of each class are determined with statistical classification techniques. The resulting spectral signatures are then used to classify each pixel of the satellite imagery for the study area into one of our land cover and use classes. This method typically misclassifies some pixels so we assess the accuracy of each mapped class by comparing the known land cover or use class type with an independent set of aerial photo samples and then with cover classes predicted through the satellite data classification.

As mentioned above, rural homes are too small to show up on Landsat imagery. Masters Student, Patty Hernandez Gude contacted each of the 20 counties of GYE and asked them for access to well permit or tax assessor data. Some counties sent digital copies; others invited us to come to their offices and photocopy their paper summaries of these data. A few counties had not compiled their data into data summaries. In these cases, Patty spent many days in the basements of county assessor offices going through individual home records one at a time. After months of work, we had a complete record at a resolution of one square mile of all rural

Confidence Intervals

In ecological studies, it is often not feasible to measure all individuals in a population or all places across an ecosystem. Instead, we strive to collect samples from a population or system and use these samples to estimate an attribute of the populations, such as density of a particular species or number of species in the ecosystem. These estimates usually differ from the true population attribute because the samples are an incomplete representation of the full population. We use statistical approaches to estimate how much our estimate from samples is likely to differ from the true population attribute.

One statistical measure is the confidence interval. A confidence interval is a range of values within which the true population mean occurs with a particular probability. For example, each time we do a bird point count in a particular habitat type, we tally a slightly different number of bird species, generally between about 10 and 20 species in aspen habitat, for example. We do many point counts in each habitat type, generally more than 100, and estimate the average or mean number of species encountered in each sample. This mean for aspen is about 17 species. We then use the variation among samples in the data to estimate a confidence interval for this estimate. We found that the 95% confidence interval for bird species richness in aspen was 15.5 - 19.5 species. This means that there is a 95% probability that the true population mean lies within this interval of species richness. The narrower the confidence interval, the higher our level of certainty that our sample estimates are good descriptors of the overall population.

homes across the ecosystem and the year they were built since 1860. The accuracy of these records was quantified by comparison with aerial photographs.

We found that the dominant change in land use across GYE was from natural and agricultural land uses to urban and exurban development. Developed land increased faster than the rate of population growth - while the GYE experienced an increase in population of 58% from 1970 to 1999, there was a 350% increase in the area of rural lands supporting exurban development. While GYE is thought to be a large wilderness landscape, we found that some 11% of the total land area of the GYE and 43% of the unprotected land area have been converted to urban, exurban, and cropland uses.

The locations of the rural homes were surprising. Yes, fertile valleys like the Gallatin Valley in Montana and the Snake River Plains in Idaho were covered with rural homes, just as Bruce and I had seen from the top of Hummingbird Peak. Unexpectedly, however, homes ringed virtually the entire public land boundary around the entire GYE. Some of these places, such as the Wind River drainage in Wyoming are very remote from cities or airports. Virtually every river or stream draining the Yellowstone Plateau is lined with rural homes. Despite the low density of people in GYE and the immensity of the wildlands, a surprising amount of the lower elevation habitats contain rural homes. Among the aspen and willow on private lands in the GYE, only 51% are free from intense human land use when defined as more than 1.6 km from agriculture, rural homes, or urban areas. Only 11% of streamsides are not near homes, farms, or cities.

These maps revealed that rural homeowners often selected the same habitats as bird species. Bird hotspots and rural homes both are both concentrated in the small percentage of the landscape that is on fertile soils at lower elevations with warmer temperatures and near riparian deciduous forests. Thus, while rural homes cover a relatively small percentage of the lands of Greater Yellowstone, we learned that they are concentrated in the key areas of the landscape that are important for native species.

Effects of Rural Homes on Biodiversity

Ecologists in the western US have long studied the ecological effects of human land uses such as livestock grazing, crop farming, mining, and timber production. The effects of rural homes on ecosystems, however, were virtually unstudied in the Yellowstone area when we began this work in the early 1990's. Perhaps ecologists presumed that the influence of scattered homes across the landscape was small compared to logging, mining, and livestock grazing that can be more conspicuous.

With little local research to draw on, we decided to read all available studies from other places and synthesize their results into a general model of rural home effects on ecosystems. This model included four general ways by which rural homes can influence biodiversity: altering or destroying natural habitats; altering ecological processes such as fire and flooding; favoring some weedy or predatory species that negatively impact other native species; and disturbance or even death of wildlife by pets and homeowners.

The net effect of these four mechanisms is that wildlife communities tend to change in close proximity to rural homes or in areas with increasing housing density. Areas associated with homes have weedy plants like spotted knapweed and dandelions and more midsized predators like crows, jays, skunks, and raccoons. Consequently, native prey species like cup-nesting birds are reduced near rural homes due to the abundance of meso predators. Large predators are also reduced with increasing home density due to displacement by dogs, road kill, and direct human persecution.

This synthesis of previous studies led us to hypothesize that birds nesting in hotspot habitats may have reduced reproduction because of the effects of nearby rural homes. Because the mesopredators like foxes, skunks, and coyotes favor rural homesites, they are also abundant near livestock and crop agriculture. We quantified from our land use data sets the density of rural homes and area of croplands within 6 km of each of our bird study stands. To get data on livestock densities, we identified pastures on aerial photos and then contacted landowners to get estimates of stocking densities in proximity to the bird sampling stands.

Jay Rotella, who led the work on bird reproduction and survival, used statistical techniques called model selection to quantify the relative importance of natural factors such as habitat type and elevation class and, human factors including density of rural homes, livestock density, and area in croplands to bird reproductive success. Such statistical analyses require large sample sizes. Although we had substantial nest success data for 6 or more bird species, we focused the analyses on two species for which we had very large samples - yellow warblers and robins.

Yellow warblers are typical of many bird species in the study area in being highly susceptible to predators and brood parasites. The yellow warbler is a small colorful bird that nests in open, cup-like nests that are generally placed in shrubs 1-2 m above the ground. While yellow warblers are skilled at hiding these nests, predators such as red squirrels, magpies, jays, and ravens often find them. Brown-headed cowbirds also find yellow warbler nests. These cowbirds are "brood parasites", that is, they lay their eggs in the nests of other species and the host parents often raise the large noisy cowbird young at the expense of their own smaller, less aggressive chicks. The American Robin also uses open, cup-like nests. However, unlike the yellow warbler, the robin is tenacious at defending nests from predators and cowbirds. Comparing the yellow warbler and the robin allows us to determine reproductive success for a species vulnerable to the types of changes associated with more intense land use and for a species less susceptible to these human caused effects.

We found that the abundance of cowbirds and nest predators increased significantly with home density. The choice riparian woodlands on the Gallatin Floodplain are loaded with avian predators and cowbirds, partially due to the large number of rural homes and associated activity in the vicinity. Nest success for both yellow warblers and robins is related to elevation class. Lower elevations have warmer climates that allowed for early nesting, longer nesting seasons, higher nest success. For

▲ PHOTO 1.4 Reproduction of many native bird species is reduced by predation. Red squirrels are highly effective at locating nests and eating eggs or young birds. Some predators, like squirrels, are native to the region. Others, such as raccoons, are not and prey on upland game bird, waterfowl and other ground nesting bird eggs and young. (Landscape Biodiversity Lab, MSU)

the Robin (less susceptible to cowbirds and predators), nest success is high enough (60-70%) to offset estimated mortality and thus the population is predicted to be stable or increasing; human habitation measured as home density seems to have no detrimental effect. Nest success for Yellow warblers, in contrast, is only 20-40%, largely due to the effects of cowbirds, who parasitized some 44% of the yellow warbler nests. For these birds, home density is negatively related to yellow warbler nest success. The estimated nest success was too low to offset mortality, thus the population in the sampled habitats was predicted to be declining. Contrasting undisturbed land we identified as hotspots, we can view highly disturbed lands as population sinks.

The last step in this analysis was to use the findings from the sampled stands to project yellow warbler population growth over the entire study area using a spatially explicit simulation model. In the model, yellow warbler abundance varied across the landscape based on habitat type, elevation, soil type, and net primary productivity. Nest success varied based on length of breeding season, habitat type, and density of rural homes. We found that under current conditions population growth was negative at higher elevations such as in YNP due to short breeding seasons and that population growth was negative in low elevation hotspots due to the effects of rural homes. Thus, the models predicted that the entire study area was a population "sink" for yellow warbler and the

▲ PHOTO 1.5 Homes in the rural countryside are spectacular for their scenic location and intimacy to nature. Unfortunately, development of the rural wildland interface can have regrettable consequences. Homes and roads may displace wildlife, be vectors for nonnative species, and add considerable political and economic pressure to natural forest functions such as wildfire. This fire threatened several dozen homes near Bozeman, Montana. (Andy Hansen, MSU)

population cannot support itself without emigrants from other locations.

Finally, we simulated "natural" conditions across the study area by statistically removing the rural home effect. The results predicted that the low elevation hotspots had strong positive population growth and that they offset the negative growth at higher elevations. In total, the results were consistent with the possibility that low elevation habitats, largely on private lands, have traditionally been population source areas that maintained populations at higher elevations, such as in Yellowstone National Park, that are limited by harsher climate and poorer soils. Rural home development on low elevation private lands can favor avian predators and brood parasites, and convert population source areas for species like yellow warblers to population sinks. The lesson is that warbler subpopulations in Yellowstone National Park risk extinction if we lose the low elevation populations that "subsidize" park populations. This is one example of how the seemingly harmless rural homes scattered around the edge of the public lands of GYE may have negative effects on wildlife inside the protected lands.

Strategies for Conservation

Knowing how human land use like rural homes influences biodiversity allow people to develop strategies to minimize negative impacts. Such knowledge can be used to help rural homeowners to live more lightly on the land. We have worked with the Sonoran Institute and other conservation organizations to develop a brochure for homeowners providing guidance on how to manage pets, livestock, weeds, and many other factors to reduce negative ecological impacts. The brochure has been very well received and many homeowners can pride themselves on the ways they have reduced conflicts with wildlife on their lands.

Policy makers and land managers can also benefit from knowing about the complexity of interrelationships between biodiversity and land use. For example, many decision makers would like to know which parcels of private lands have high value for biodiversity and are likely to be developed in the near future. Such lands become high priorities for conservation easements and other incentive-based approaches for protecting private lands of high conservation value.

Patty Hernandez-Gude projected future rural home distribution across the GYE and assessed impacts of various measures of biodiversity. We analyzed the ecological and socioeconomic correlates with past growth in rural home density to parameterize a computer model to predict future growth out to 2020. We also compiled data on some 11 measures of biodiversity relating to habitats of individual species, communities of species (e.g., bird hotspots), and integrated indices of biodiversity that included combinations of individual measures. We overlaid projected home density on the biodiversity maps. This allowed us to determine the parcels of land that are most likely to have rural home construction and that that have the highest value for biodiversity. We provided the resulting maps to land trusts and other groups that are active in working with land owners to develop conservation easements. As a result, some of the most important and vulnerable private lands around GYE are now in a protected status.

Conclusion

After some 15 years of research on rural homes in the GYE, we are beginning to understand where they occur in the landscape, rates of increase, impacts on biodiversity, and ways for home owners and policy makers to minimize negative influences. Although the GYE is very large and mostly wild, we have developed a set of methods that help to understand the complex interactions of people and wildlife in the area. The results of this set of studies of biodiversity and rural homes have provided important information on how to manage GYE to sustain both native species and the human communities in the area.

These methods have also evolved into a more general approach for monitoring and analyzing national parks and their surrounding greater ecosystems that we are now applying across the United States. The approach integrates field data, remote sensing, and simulation modeling to "take the pulse" of the system, analyze trends, and recommend management actions.

Key steps of the approach are as follows:

- Identify the key biotic resources of interest (e.g., native species) and the natural and human factors that influence them;
- Delineate the boundaries of the surrounding greater ecosystem on which the national park is dependent;
- Use remote sensing and other methods to monitor change in the key resources and in the drivers;
- Analyze these data to identify trends past to present, and likely trends into the future that may push the key resources over negative thresholds of change;
- Deliver the resulting data, maps, and knowledge to park managers so that they can include this information when they make decision about park management.

▲ PHOTO 1.6 The most powerful threats to biodiversity in places like the Greater Yellowstone Ecosystem are habitat loss, degradation, and fragmentation. Roads and rural residential development convert natural landscapes to alternative, sometimes damaging, land uses. Careful planning and conservation can mitigate some changes but communities must also cultivate a political will to preserve environmental amenities. (Andy Hansen, MSU)

These methods are among several being used by the new National Park Service Inventory and Monitoring Program. This program is tracking the health of all the national parks and national monuments across the US and providing key information to park managers to keep them ecologically healthy. While Greater Yellowstone is a large, complex system, it is but one piece of a national park structure. The same methods that have helped us understand and manage Greater Yellowstone have promise for working across the entire parks network. Jerry, Bruce, and I have many slopes to ski before we fully understand this complex ecosystem.

Chapter 2

Mapping the Last Frontier in Yellowstone National Park: Yellowstone Lake

Chapter 2

Despite a history of doing earth science and geologic mapping in the Park since the 1870's, no one had created a geologic map of Yellowstone Lake. The Hayden Expedition explored and mapped the shoreline in 1871 and produced the first map of the floor of the lake. Others followed but it was not until 1999 when USGS geologists Lisa Morgan and Pat Shanks with a multi-disciplinary research group began the first comprehensive mapping studies. Their aim was to produce the most detailed maps modern technology allowed. To do so they employed an array of high tech sensors attached to a National Park Service boat and an underwater rover similar to the one that explored the Titanic. They processed five years of data to produce maps of incredible resolution and revealed a host of underwater features and geologic wonders.

Yellowstone Lake is the product of glacial processes that scoured the Yellowstone River valley and, dramatic volcanic and hydrothermal activity that originates deep in the earth's crust. Imagine an eruption 640,000 years ago where Yellowstone is today. Over 1,000 cubic kilometers of pyroclastic flows - a mixture of hot, dry ash, rock fragments, and hot gases blasted across an area nearly 7500 square kilometers. The resulting caldera (volcanic depression) formed the core of YNP and the northern portion of Yellowstone Lake. The maps produced by Morgan and Shanks' team show that the lake itself holds all manner of dramatic thermal features, including -hydrothermal vents, large circular craters due to steam explosions, hydrothermal domes, and subterranean faults and fissures. The vents are sources of various chemicals - notably mercury, a substance with well-known toxic properties at high levels that has turned out to be useful to ongoing ecological studies on the Yellowstone grizzly bear.

The maps show a complex landscape of past geothermal events, landslides, faults, changing shorelines and evidence of potentially hazardous seismic activity; it will be decades before geologists fully understand these features. The true value of the maps may be the capacity to meld the science of the hidden geology of Yellowstone Lake with our understanding of the terrestrial landforms that fascinate scientists and visitors alike.

J. Johnson

Yellowstone Lake (Thermal Biology Institute, MSU)

Chapter 2

Mapping the Last Frontier in Yellowstone National Park: Yellowstone Lake

Lisa A. Morgan and W.C. Pat Shanks

Lisa A. Morgan, U.S. Geological Survey, 973 Federal Center, P.O. Box 25046, Denver, CO 80225-0046; Email: lmorgan@usgs.gov

W.C. Pat Shanks, U.S. Geological Survey, 973 Federal Center, P.O. Box 25046, Denver, CO 80225-0046; Email: pshanks@usgs.gov

For more information on science in Yellowstone Lake: http://volcanoes.usgs.gov/yvo/new.html

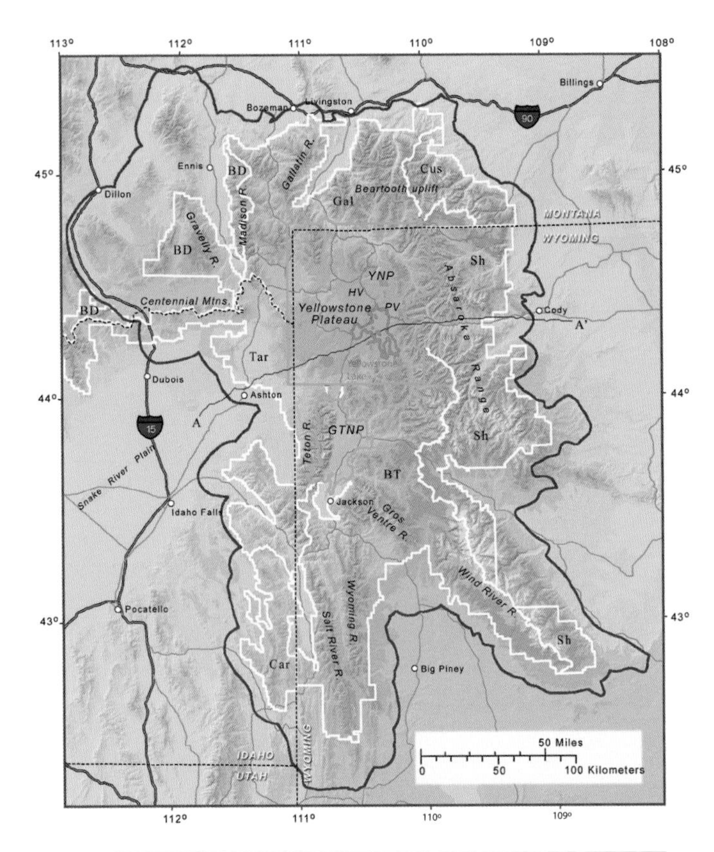

▲ FIGURE 2.1 Index map of the greater Yellowstone geoecosystem (GYE) (outlined in red) with Yellowstone Lake in the center of the geoecosystem. (Ken Pierce, USGS)

Yellowstone Lake is the centerpiece of the Greater Yellowstone Ecosystem. The lake is the largest high-altitude (>2134 m in elevation) lake in North America, covering 341 km^2 and carrying 16.54 km^3 of water.

The headwaters of the Yellowstone River, Yellowstone Lake figured prominently in determining the course of the 1871 Hayden survey expedition of the Yellowstone region. A primary scientific interest of Hayden was watersheds; thus, a principal goal of the Yellowstone survey was to reach the headwaters of the Yellowstone River and map Yellowstone Lake. The latter goal was accomplished by survey party members Henry Elliott and others who collected in 24 days over 300 soundings using triangulation for navigation and produced the first bathymetric map of Yellowstone Lake. A bathymetric map displays the ocean or lake floor terrain as contour lines called depth contours or isobaths.

Since the Hayden survey, several other maps of Yellowstone Lake have been created, each using more advanced technology than the previous bathymetric map. Each map represents a step forward in more accurately defining the lake boundary and lake floor morphology and has improved understanding of what is present in the lake and how it may have formed. Up until 1990, however, accurate navigational systems were not readily available and map resolution was relatively low. Thus, details of the lake floor could not be accurately defined. In the late-1990's, a collaborative effort between various organizations to study the chemistry of fluids from hydrothermal vents on the floor of Yellowstone Lake resulted in the recognition that the maps available at the time were inadequate in identifying the locations of individual hydrothermal vents or fields. In 1999, the U.S. Geological Survey (USGS) in collaboration with the National Park Service at Yellowstone National Park began a 5-year, high-resolution mapping survey of Yellowstone Lake. This survey would be the first to utilize and integrate new technologies involving differential GPS for navigation, multi-beam swath sonar, and high-resolution sub-bottom seismic-reflection profiling. The data sets were complimented by sampling and photographic documentation using a submersible remotely operated vehicle (ROV). The final map produced a data set with navigational accuracy of <1 m, multi-beam swath sonar data accurate to <1 m both horizontally and vertically, and seismic reflection profiles of the upper 25 m of the lake floor to a vertical resolution of 10 cm and horizontal resolution of 3 m (at 10 m water depth) to 13 m (at 50 m water depth). What this means in practical terms is that features as small as one meter could be observed with the newer technology.

The merits of such a high-resolution map of the lake were well justified. Significant advances had been made in the 1990's in developing high-resolution maps of the depths of the ocean floor; however, no high-resolution

Deep Water Exploration

The array of available technology for underwater mapping and geophysical surveys has changed dramatically over the last couple decades. Various sonars and acoustic sensors, magnetometers, and free-swimming, tethered rovers allow scientists to see highly detailed, three-dimensional images of the floor of large bodies of water like Yellowstone Lake. The resolution can be as accurate as 10s of centimeters. Below are short descriptions of some of the methods used to produce the highly detailed maps of the Lake.

▶ *SEISMIC-REFLECTION PROFILING.* A variety of sources, including explosives, pneumatic air guns, and high-frequency "chirp" systems, are used in seismic-reflection surveys. A "chirp" sonar system that transmits computer-generated pulses of acoustic energy was used in Yellowstone Lake. The sound travels down through the water and penetrates into the layers of sediments and rocks on the lake floor directly beneath the boat. Some of this sound reflects (echoes) off the layers, and travels back up to the surface where it is recorded by a hydrophone. The return time and intensity of the returning energy helps researchers know the structure of the lake floor and underlying layers. A series of pulses over time establishes a profile of the lake bottom as the boat drives along a prescribed course. The sonar pulse is harmless to fish and the surrounding environment. High resolution seismic reflections surveys are particularly effective for assessing the sedimentary layering and other features in the upper 10-30 m of sub-bottom materials. Sub-bottom seismic-reflection profiling utilized an EdgeTech SB-216S, which sweeps a frequency range from 2 to10 kHz and has a beam footprint that spreads over an angle of 15-20°.

▶ *HIGH-RESOLUTION MULTI-BEAM SWATH SONAR.* A hull-mounted transducer sends out 126 energy beams in a 150° fan-like array, somewhat like a peacock's tail. The energy is bounced back to collectors that interpret the return signal to detail the bottom bathymetry (underwater topography). The fan-like array is oriented perpendicular to the direction the boat travels, so it maps a swath that is about 8-times water depth in width. The boat is driven on a back-and-forth course mostly in a 200-m north-south spacing so the swaths overlap, like mowing a lawn, giving continuous coverage of the bottom as the survey progresses. Swath mapping differs from seismic reflection profiling in focusing on obtaining a continuous map of the shape of the bottom. Higher frequency sound is generally used, which gives a sharp reflection off the bottom, but does not penetrate significantly into the sub-bottom sediments.

▶ *AEROMAGNETIC SURVEY.* Airborne geophysical surveys are carried out using a magnetometer aboard or towed behind an aircraft. The principle is similar to a hand-held metal detector, but allows much larger areas of the Earth's surface to be covered quickly by aerial reconnaissance. The aircraft typically flies in a grid pattern with height and line spacing determining the resolution of the data. The magnetometer records tiny variations in the intensity of the earth's magnetic field due to subsurface structures and the amount of magnetic minerals in the Earth's crust. Aerial surveys are fast and magnetic properties can be used to detect different rocks types and different degrees of hydrothermal alteration.

▶ *ROVERS/ORV.* A wide array of rovers (Remotely Operated Vehicle) are available to researchers. Some are manned, others are robots tethered to the mother ship, and a new class of autonomous underwater vehicles (AUVs) is untethered and can be programed to carry out surveys. All of these vehicles have their own propulsion systems and are able to maneuver and investigate features of interest. In Yellowstone, we used a machine that relays live video to the surface and can take water, sediment, rock, and biota samples. One advantage of robots is that interesting features can be closely investigated without the risk associated with human dives to several hundred feet. They can be easily and safely controlled from the surface ship.

▲ PHOTOS 2.1 & 2.2 The National Park Service research vessel, the *RV Cutthroat*, was used to collect the bathymetric and seismic data as well as provide the platform for the submersible remotely operated vehicle (ROV). The boat is used by researchers mapping underwater features as well as taking biological samples of lake flora and fauna. (Big Sky Institute, MSU)

bathymetric map existed for Yellowstone Lake. In fact, because of the low resolution of previous lake maps, details of the shape and dimensions of the landforms and geology of the lake floor were poorly known. Yellowstone Lake was left as a large blue hole surrounded by detailed topographic and geologic maps on land. With no geologic map of Yellowstone Lake available, basic information such as how the lake formed, how the lake fits in with its surrounding geology, and what potential hazards or resources were present in the lake were poorly known. Furthermore, a high-resolution bathymetric map enabled discovery of several major new features and important details for a multitude of smaller features. Coupling the collection of high-resolution multi-beam swath sonar imaging of the lake floor with high-resolution seismic reflection profiling gave the survey the first comprehensive coverage of ~30-m-thick sub-bottom slices into the lake floor. These two new surveys complimented data collected from a high-resolution aeromagnetic survey of Yellowstone National Park (YNP) by the USGS in 1996. Integration of these distinct but complimentary data sets resulted in the USGS producing high-resolution bathymetric and shallow seismic maps of Yellowstone Lake as well as the first geologic map that provides a basic framework in which to identify potential hazards in the lake and to effectively manage the various resources present.

At the same time, the National Park Service was facing a significant challenge in managing certain natural resources in Yellowstone Lake. In 1994, aggressive fish-eating lake trout were discovered in the lake. Each lake trout in Yellowstone Lake is estimated to consume about 60 native cutthroat trout annually, rapidly decimating the cutthroat population. Lake trout live their entire life cycle within Yellowstone Lake. In contrast, spawning cutthroat trout annually swim up one of the 141 tributaries that drain into Yellowstone Lake where they become an important food source for the grizzly bear, bald eagle, osprey, and river otter. Without the availability of cutthroat trout as an integral part of their diet, these species are compromised creating a ripple effect through the entire ecosystem and adversely affecting an area much larger than Yellowstone Lake. The National Park Service was interested in a high-resolution map of the lake to help identify areas where lake trout might spawn, then focusing gill-netting operations on those areas to reduce the population.

The five year collaborative survey of Yellowstone Lake by the USGS and the NPS cost about $600,000 for outside contract services alone, supported through a combination of mostly public funds with some private contributions. Multiple divisions and programs in the USGS (including the Mineral Resources Program, Volcano Hazards Program, Climate History Program, Northern Rocky Mountain Science Center, Biologic Resources Division, Geologic Division Venture Capital Program, and the Central Region Office of the Director) contributed to this effort as did the NPS (Yellowstone

National Park), the Yellowstone Foundation, and the Yellowstone Association.

Previous Geological and Geophysical Studies

A variety of detailed field studies in YNP have been conducted by the USGS since the Hayden survey in 1871. In the late 1960's, in an effort to improve our understanding of young volcanic terrains and how these young volcanic terrains may be applicable to lunar and planetary geology, NASA supported a major mapping effort by the USGS to map the entire park. Detailed field mapping of the young (<2.1 million year) volcanic rocks of YNP was completed, and set the stage for understanding the geologic framework of Yellowstone Lake.

A 1977 a heat flow map of the lake showed that the Mary Bay area in the northern basin of Yellowstone Lake had the highest heat flow values in the lake and depth to magma may be five to eight kilometers, some of the shallowest magma in the park.

In 1996, the USGS collected high-resolution aeromagnetic data for the entire area of YNP. This survey was flown along closely spaced (every 400 m), north-south trending flight lines at low flight elevations (draped at <350 m above existing terrain) allowing resolution of low amplitude, short-wavelength magnetic anomalies that could be used to resolve features at scales useful for mapping individual geologic units, faults, and areas of alteration. This mapping replaced the previous lower resolution aeromagnetic maps that were flown in 1973.

Methodologies

A multi-disciplinary approach was used in mapping Yellowstone Lake. Existing data included detailed geologic maps of YNP and the surrounding on-land areas around the lake as well as high-resolution aeromagnetic data of YNP. To this set, we added two sonar surveys (swath multi-beam sonar and seismic reflection profiling). Navigation for the surveys used a real-time, differentially corrected global positioning system (GPS), which resulted in a location accuracy for the survey vessel of <1 m. Features identified in the seismic and bathymetric surveys were verified using a submersible remotely operated vehicle (ROV) to photographically document the lake floor and sample for solids and fluids. The solid and fluid samples were then analyzed for trace element chemistry and isotopes. Several samples were radiometrically dated and probed with a scanning electron microscope.

Bathymetric surveys of Yellowstone Lake occurred in four campaigns between 1999 through 2002 and utilized a Sea-Beam 180 kHz instrument with a depth resolution of <1% water depth. Water depth in Yellowstone Lake varied from ~4 m to 133 m in the areas surveyed. The multi-beam instrument used has 126 beams arrayed over an angle of 150° angle to map a swath width of 7.4 times water depth. Data were collected primarily along north-south lines spaced approximately every 200 m; in areas with shallow water, the spacing between collection lines was closer. Multiple east-west tie lines were collected for increased accuracy. Over 240,000,000 soundings were collected to produce the first high-resolution, continuous and overlapping coverage of the bathymetry of Yellowstone Lake.

Coupled with the bathymetric survey, sub-bottom seismic reflection profiling utilized an EdgeTech SB-216S, which sweeps a frequency range of 2-15kHz, and has a swath angle of 15-20°. Both the swath unit transducer and the sub-bottom unit were rigidly mounted to the transom of an 8-m long aluminum boat used for survey purposes. Approximately 2400 linear km of high-resolution seismic reflection data, which penetrated the upper 25 m of the lake bottom, were collected between 1999 and 2003.

The Eastern Oceanics submersible ROV is a small vehicle (~1.5 m x 1 m x 1 m) attached to the vessel (RV Cutthroat) with a 200-m tether, which was operated in tandem with these surveys. The ROV provides live videographic coverage and remote control of cameras and sampling equipment. The ROV has full-depth rating of 300 m and is capable of measuring temperature, conductivity, and depth, and can retrieve hydrothermal vent fluid samples and solid samples up to 40 cm long.

Results and Discoveries

CHRONOLOGY OF DISCOVERIES FROM THE 1999-2003 CAMPAIGNS OF YELLOWSTONE LAKE

Mapping of Yellowstone Lake by the USGS in collaboration with NPS began in 1999 and concluded in 2003. Initial efforts started in the northern basin in 1999 and continued in 2000 in West Thumb basin. The central basin followed in 2001; all three surveys between 1999 through 2001 collected multi-beam swath sonar and sub-bottom seismic-reflection data contemporaneously. In 2002, multi-beam swath sonar data was collected for the Southeast, South, and Flat Mountains Arms in the southern reaches of the lake; additionally a small feature in the northern basin was remapped. In 2003, seismic-reflection data was collected from selected lines in all arms of the lake as well as certain other areas of interest the lake. In 2003, a significant effort was made into sample and photographically document the lake floor with the submersible ROV.

With subsequent years, our findings grew increasingly interesting and complex, building on knowledge of previous years. Our ROV surveys ranged from inventorying and identifying geographical features to discovering new and exciting ecological ties between the hydrothermal vents, the lake fishery, and their predators.

The primary discoveries from the mapping of Yellowstone Lake in 1999 were a broad range of features related to hydrothermal activity. These included individual hydrothermal vents, sets of vents, linear fractures (many lined with hydrothermal vents), large hydrothermal explosion craters, hydrothermal domes, and siliceous spires related to hydrothermal vents. The lake was found to be a complex network of hydrothermal features in a large previously unknown thermal basin. Other discoveries from the northern basin survey included identification of landslide deposits at several locations and the young, active Lake Hotel graben, a geological two-sided fault whose precise location was determined in the new survey.

In 2000, mapping focused in the West Thumb basin.

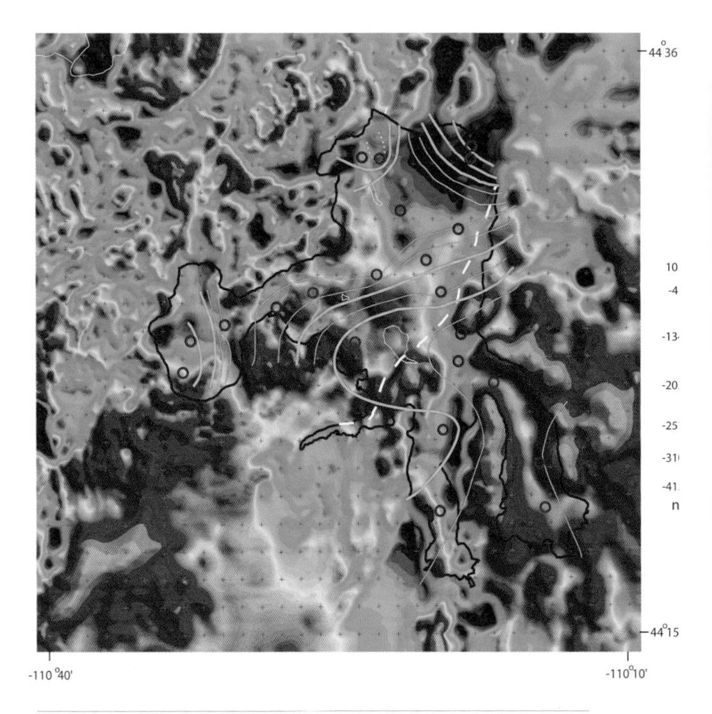

▲ FIGURE 2.2 The rainbow-colored shaded relief bathymetric map of Yellowstone indicating depth throughout the lake is based on multibeam sonar imaging. Yellowstone Lake is outlined in black, the caldera margin is represented by dashed thin white line. The most recent bathymetric and seismic maps reveal new faults, hot springs and craters beneath Yellowstone Lake. (Lisa Morgan, USGS)

Page 26 FIGURE 2.3 High resolution reduced to pole aeromagnetic map of Yellowstone Lake at a 100 m grid, magnetic inclination = 69.8, magnetic declination = 14.3. (Lisa Morgan, USGS)

The basin erupted less than 200,00 year ago within the Yellowstone caldera. The high concentration of hydrothermal activity makes it an area of scientific interest. Several previously unknown large hydrothermal vent fields and one large hydrothermal explosion crater were mapped. Another major discovery in 2000 was that rhyolitic lava flows were emplaced into the lake-filled West Thumb caldera created by the eruption of the tuff of Bluff Point ~180,000-190,000 years ago. Using the newly acquired bathymetry from West Thumb basin and draping it on top of the high-resolution magnetic intensity map of the same area, we noted a strong correlation between the two data sets. On land, sharp changes in magnetic amplitude were strongly associated with geologically mapped edges of rhyolitic lava flows; topography also reflects these changes.

Coupling the newly acquired bathymetric data with the aeromagnetic data showed similar relationships in the lake, and prompted our proposal that rhyolitic lava flows were also in Yellowstone Lake. These flows contributed significantly to shaping the northern two-thirds of the lake floor.

Newly acquired seismic reflection profiles across the West Thumb basin showed a pronounced high amplitude reflector identified at about 9 m depth along a north-south transect in the northern part of West Thumb basin. This is interpreted as the top of a rhyolitic lava flow covered by laminated glaciolacustrine sediments (glacially-derived sediments deposited in the lake). Older geologic maps of the northern and western sides of the lake show exposures of similar post-Yellowstone-caldera lava flows adjacent to the lake. Based on this spectrum of data, we identified and mapped rhyolitic lava flows as present in the northern, West Thumb, and central basins of Yellowstone Lake. In fact, these lava flows are the major contributors to the overall morphologies present in the northern two-thirds of the lake. Prior to this, most researchers thought the landforms in these basins of the lake were related to the advance, deposition, and retreat of glaciers. In fact, glacial debris is mapped on the various islands within Yellowstone Lake; however, based on the 2000 survey data, these thinner glaciolacustrine (glacier deposited) units cover the larger, more impressive forms created by the rhyolitic lava flows in the northern two-thirds of the lake.

What all this means is that the creation of Yellowstone Lake is a combination rarely seen in high mountain environments. Normally, lakes in environments like Yellowstone are the result of being gouged out by glaciers as they moved across the landscape and Yellowstone is no exception. However, the northern two-thirds of the lake has been shaped by a combination of dramatic geologic processes including Yellowstone caldera formation, emplacement of rhyolitic lava flows, glacial advances and retreats, hydrothermal explosions, and relatively rapid changes in lake level due to deformation (inflation and subsidence) of the Yellowstone caldera.

The opportunity to map the expansive central basin of Yellowstone Lake began in 2001 and more discoveries ensued. Our mapping continued to identify multiple hydrothermal vents, several large hydrothermal explosion craters, small landslide deposits, and large areas covered by rhyolitic lava flows. For the first time, the precise location of the topographic margin of the 640,000-year-old Yellowstone caldera was defined. In addition, the linkage of the young and active north-south Eagle Bay fault system with the newly mapped fissure system located due west of Stevenson Island to the north, continuing north to the Lake Hotel graben on the northern outlet of the lake was made. On the eastern side of the basin a large (1 km diameter) detached block of Tertiary volcanic rock was discovered on the lake floor. In each of these new discoveries, the interpretation of the data was supported by the bathymetry coupled with the magnetic intensity map and/or data from the seismic reflection profiles, which thereby strengthened the interpretation. The submersible ROV provided the "ground truth" in the form of high quality photographic images and solid and fluid samples.

Efforts in 2002, the final year for collecting multi-beam swath sonar data had a two-fold focus. One objective was to complete the mapping of Yellowstone Lake and finish the southern-most parts of the lake, including the Flat Mountain, South, and Southeast Arms. The second objective was to re-map an area in the northern basin of the lake, which has been interpreted as a large (~700 m in diameter) and active hydrothermal dome. We wanted to determine if any movement could be detected by comparing newly collected data with the 1999 survey of the same area. No differential movement was detected (within the 60 cm margin of error) in the four years between 1999 and 2002.

Mapping the Arms in 2002 revealed a glaciated and faulted landscape bounded on its north by the topographic margin of the Yellowstone caldera. The bathymetry of the lake floor, especially in the Southeast Arm, shows many glacial meltwater features and stagnant ice block features. Similar features, such as Alder Lake, are mapped on land at the Promontory. Both the Southeast and South Arms are bounded by active

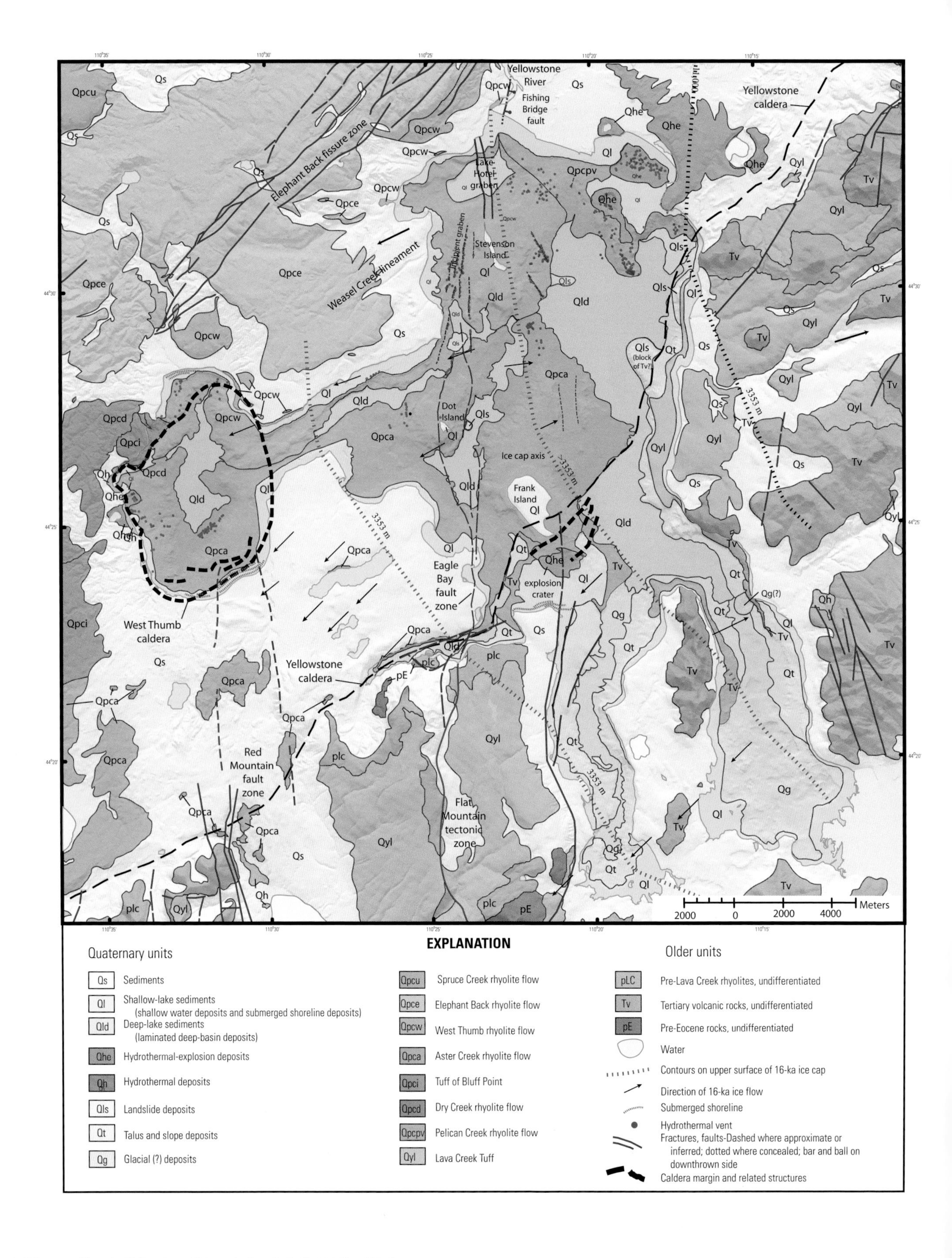

Yellowstone River
Fishing Bridge fault
Yellowstone caldera
Elephant Back fissure zone
Lake Hotel graben
Incipient graben
Stevenson Island
Weasel Creek lineament
Dot Island
Ice cap axis
3353 m
Frank Island
Eagle Bay fault zone
explosion crater
West Thumb caldera
Red Mountain fault zone
Flat Mountain tectonic zone
Qls (block of Tv?)
Qg(?)
2000 0 2000 4000 Meters
EXPLANATION
Quaternary units
Qs Sediments
Ql Shallow-lake sediments (shallow water deposits and submerged shoreline deposits)
Qld Deep-lake sediments (laminated deep-basin deposits)
Qhe Hydrothermal-explosion deposits
Qh Hydrothermal deposits
Qls Landslide deposits
Qt Talus and slope deposits
Qg Glacial (?) deposits
Qpcu Spruce Creek rhyolite flow
Qpce Elephant Back rhyolite flow
Qpcw West Thumb rhyolite flow
Qpca Aster Creek rhyolite flow
Qpci Tuff of Bluff Point
Qpcd Dry Creek rhyolite flow
Qpcpv Pelican Creek rhyolite flow
Qyl Lava Creek Tuff
Older units
pLC Pre-Lava Creek rhyolites, undifferentiated
Tv Tertiary volcanic rocks, undifferentiated
pE Pre-Eocene rocks, undifferentiated
Water
Contours on upper surface of 16-ka ice cap
Direction of 16-ka ice flow
Submerged shoreline
Hydrothermal vent
Fractures, faults-Dashed where approximate or inferred; dotted where concealed; bar and ball on downthrown side
Caldera margin and related structures

normal faults and, in tandem with glaciation, contribute to the overall form of the southern valleys.

The topographic margin of the Yellowstone caldera was determined to enter the western end of Flat Mountain Arm and continue along the channel of the arm. The caldera margin then enters and continues east-northeastward across the central lake basin. From the eastern shore of Frank Island, the caldera margin cuts northwest and emerges between the older Tertiary volcanic Lake Butte and the younger Turbid Lake hydrothermal explosion crater.

The campaign in 2003 was devoted to 1) sampling hydrothermal vent fluids and solids from the floor of Yellowstone Lake using a submersible ROV and 2) collecting sub-bottom seismic-reflection data for areas of special interest in the South and Southeast Arms and Central and Northern Basins. Just south off the Rock Point area, multiple young and active faults with small displacements that penetrate the lake floor were discovered in seismic profiles.

Significant Findings from Mapping of Yellowstone Lake

FIRST GEOLOGIC MAP OF YELLOWSTONE LAKE

Until the recent effort to map Yellowstone Lake, it remained a final frontier in the midst of detailed geologic mapping on land in Yellowstone National Park. As a result of the recent detailed mapping, we understand that the processes forming Yellowstone Lake can be separated into two major areas of the lake: 1) the northern two-thirds of the lake, including West Thumb basin, was formed primarily by volcanic and later hydrothermal processes; and 2) the southern third was shaped primarily by glacial and alluvial processes. Tectonic forces have and continue to influence all areas of the lake.

Yellowstone Lake is the centerpiece of the Yellowstone geoecosystem and has been shaped by the main geologic forces that have created the Yellowstone landscape. Formation of the lake probably predates formation of the 640,000-year-old Yellowstone caldera and may extend as far back as 2.05 million years ago when the Huckleberry Ridge caldera erupted. The eastern edge of that caldera is estimated to extend north-south through the central part of the lake. Physical evidence for formation of a lake can be inferred as far back as 640,000 years ago when the Yellowstone caldera erupted and formed a topographic basin bounded by a topographic rim on its southeastern part, now preserved in Yellowstone Lake. Shortly thereafter, large, previously unmapped rhyolitic lava flows were emplaced in the northern basin area.

Approximately 190,000 to 180,000 years ago, another caldera-forming eruption, this time much smaller and within the larger Yellowstone caldera, created the West Thumb basin formed from the eruption of the tuff of Bluff Point, which is now exposed along the east side of West Thumb basin. Around 154,000 to 150,000 years ago, several large rhyolitic lava flows were emplaced, first into the southern half of West Thumb caldera as the Aster Creek rhyolite flow and shortly after, into the northern half of West Thumb caldera as the West Thumb rhyolite flow. Both of these were massive lava flows from outside the West Thumb crater toward the east into the northern and central basins.

Glacial activity also played a dominant role in formation of Yellowstone Lake. Major and minor glacial advances and retreats occurred from ~180.000 to 140,000 years ago and from ~70,000 to 16,000 years ago. A north-south-trending axis of ice over 3000 feet thick is estimated to have covered the central basin before receding about 16,000 years ago.

Starting around 13,000 years ago, large (>100 m in diameter) hydrothermal explosion craters began to form in and around Yellowstone Lake; at least four to six large explosion craters have been mapped in the lake. At least six other large, post-glacial hydrothermal explosion craters have been mapped on land around or near Yellowstone Lake, making this area noteworthy for having the highest concentration of large hydrothermal explosion craters or domes in the park. Overlapping this time frame, various lake shore lines have been mapped along the perimeter of Yellowstone Lake, which reflect the inflation and deflation of the Yellowstone caldera. Faulting associated with the mostly north-south trending fault segments such as the Lake Hotel graben, the Eagle Bay fault zone and the faults /seismic activity due south of West Thumb basin, has been contemporaneous with the current deformation of the caldera.

These geologic processes are reflected in the geologic map of Yellowstone Lake. Numerous large-volume, post-caldera rhyolitic lava flows, hundreds of hydrothermal vents, large hydrothermal explosion craters, fissures, faults, landslide deposits, hydrothermal domes, and submerged shoreline deposits also have been mapped. The identification of these deposits or features as landslide and hydrothermal explosion deposits, faults, and chemical dissolution craters indicate potentially hazardous events that have occurred recently in the lake and possibly could occur again.

▲ PHOTO 2.3 Images of siliceous spires mapped in Bridge Bay of Yellowstone Lake. A) Photomosaic of a small siliceous spire. Dimensions are ~5 m tall and ~3 m wide at base. B) Scanning electron photomicrograph of material from the interior of a spire showing that the spire is composed of silicified filamentous bacteria and diatoms. (David Lovalvo)

CHEMISTRY OF SUBLACUSTRINE HYDROTHERMAL VENTS

Using multi-beam swath sonar and seismic-reflection profiling, over 650 hydrothermal vents were mapped as vent craters in Yellowstone Lake. Fluids from many of these vents, as well as from 44 of the 141 tributaries entering Yellowstone Lake, have been sampled and analyzed for chemical and isotopic composition. Results show that about 10% of the total chloride flux (used as an indicator of geothermal activity) in Yellowstone National Park occurs in Yellowstone Lake. Only Lower and Upper Geyser Basins are more significant thermal basins than the Yellowstone Lake basin.

Vent fluid samples obtained by ROV were analyzed for dissolved gases, chemical constituents, and stable isotope composition (Hydrogen, Oxygen, and Sulfur). Results show that the major dissolved gases are Carbon Dioxide (CO_2) and Hydrogen Sulfide (H_2S). Stable isotope studies show that vent fluids are meteoric water (rainfall and snowmelt), while H_2S is derived from deep magmatic degassing or leaching of sulfide minerals in the underlying volcanic rocks. Geochemical studies of dissolved constituents in hydrothermal vents show that vent fluids are strongly enriched with a variety of potentially toxic elements, notably mercury, arsenic, molybdenum, and tungsten. In general, this composition is similar to other subaerial hot springs, thermal pools, and geysers in the park.

Vent fluids carry significant dissolved silica (SiO_2) and can deposit silica in conduits below the lake floor, as veins or zones of silicification in sediments, and in some cases as significant structures referred to as spires up to 8 m tall on the lake floor. Another major discovery of these studies, beautifully linking chemical and geological studies, shows that under certain conditions of mixing and cooling, vent fluids can dissolve silica in a wholesale fashion from sediments at the vent sites. This mechanism produces some of the vent craters observed by bathymetric and seismic reflection mapping.

Mercury is one of several potentially toxic metals emitted from hydrothermal vents on the floor of Yellowstone Lake and, as such, has a significant impact on its uptake into the food chain. As part of the hydrothermal studies and mapping in Yellowstone Lake project, a site called the "cutthroat jacuzzi" near West Thumb Geyser Basin was visited and sampled. This site, a low-temperature hydrothermal vent (hot spring) in about 5 m of water depth, was named by a University of Wisconsin-Milwaukee research group after noticing persistent populations of 5-20 cutthroat trout frequenting the vent area. The trout were occasionally observed feeding on bacterial mats and amphipods (small shrimp-like crustaceans) in vent waters. We immediately recognized that this might constitute a biochemical pathway introducing potentially toxic elements into the ecosystem and set about sampling cutthroat and lake trout tissue. Chemical analyses of fish muscle proceeded along with analyses of lake water, hydrothermal vent fluid, and related silica-rich sinter deposits.

Results soon indicated that both fish muscle and sinter deposits from Yellowstone Lake carry significant mercury. This resulted in a letter to then Superintendent Mike Finley about the mercury content of the fish, which was mostly below the EPA (1.0 mg/g wet weight) and World Health Organization limits (0.5 mg/g) for edible fish, with occasionally samples above these limits. The park was immediately notified and agreed to provide funding for a more thorough job of analyzing fish of different sizes and ages, and from different parts of the lake. Results indicated an average of about 0.25 mg/g mercury, which is far enough below established limits that the risk to humans from occasional consumption is small. As an interesting side note, one early proposal in the struggle to save the native cutthroat trout in Yellowstone Lake from lake trout predation considered establishing a commercial fishery in the lake to target the lake trout. With the availability of this new data, the Park Service discarded this possible plan in part because of the discovery of mercury concentration in the trout.

We then turned our attention to mercury sources and pathways in the ecosystem. Recalling our observations of trout feeding at the "cutthroat jacuzzi," we obtained and analyzed stomach contents of several lake and cutthroat trout. While large lake trout often have cutthroat in their stomachs, cutthroat trout often have abundant amphipods. In both cases, analyses showed

▲ PHOTO 2.4 When high levels of mercury were discovered in the native Yellowstone trout fishery, it was feared that grizzly bears, eagles, and other predators on the trout may be suffering. To date, there is no evidence that mercury from the deep lake vents is detrimental to the complex food web in the Park. (Big Sky Institute, MSU)

that the mercury concentrations in the stomach content were similar to, or somewhat lower than, that in the fish muscle, suggesting amphipods as the source of mercury. The mode of bioaccumulation from vent fluids to amphipods to fish has not been studied in detail, but presumably involves bacterial conversion of mercury to methyl mercury, which is incorporated into the bacterial cells and then transferred to small crustaceans during feeding.

On the other end of the food chain, we were interested in whether the mercury in cutthroat trout was accumulating in predators such as grizzly bear, osprey, eagle, and otter. Coincidentally, our colleague Bob Rye, working with Chuck Schwarz (U.S.G.S. Interagency Grizzly Bear Study Team), used stable isotopes of C, S, and N on bear hair to see if food sources could be determined using relatively new isotope techniques. We obtained hair samples collected both around tributary streams to Yellowstone Lake where the cutthroat spawn and from remote areas. Results indicated dramatically higher mercury concentration in the hair of bear that feed on cutthroat trout. Subsequent studies have shown that the spawning trout are the only source of mercury in bear foods in Yellowstone, and that the levels of mercury the bears accumulate is not detrimental to their health. Similar studies for osprey, eagle, and otter have not been carried out yet.

Policies

High-resolution geophysical mapping of Yellowstone Lake was a joint effort supported by the US Geological Survey and Yellowstone National Park from 1999-2003 and addressed key issues for both agencies. By producing a high-resolution (<1 m accuracy) bathymetric map of the lake floor with complimentary seismic reflection profiles, a detailed geologic interpretation has been developed. The new data sets were used to make an interpretive geologic map of the lake, which identified potentially hazardous deposits, features, and resources, and contributed to the volcanic and hydrothermal hazard assessment for Yellowstone National Park. The new bathymetric map allows the National Park Service Fisheries Division to focus on various management techniques in areas interpreted as prime spawning habitat for lake trout. Additional research characterizing specific sites on the lake floor have continued. The new bathymetric and geologic maps are available through several distribution centers, the authors, or on line at: http://pubs.usgs.gov/sim/2007/2973/.

Policies affected by the recent mapping of Yellowstone Lake may include: 1) new management policies toward lake trout populations in the lake; 2) the understanding of potentially hazardous events in the lake, and 3) the location and subsequent protection of rare and unusual geologic deposits. Long-standing policies regarding the exploitation of thermal species and protection of thermal features in Yellowstone are firmly in place and may now be applied to the underwater resources.

Future Research

The new bathymetric and geologic maps of Yellowstone Lake serve as the foundation for more detailed studies regarding its recent geologic history and potential hazards, a lake-wide all-taxa inventory, a baseline datum of the lake floor; and a high-resolution map to a natural laboratory allowing one to examine the interrelationships between geology and biology. Collaborative efforts are underway among microbiologists, geologists, and geochemists to conduct an all taxa inventory throughout Yellowstone Lake with an emphasis on the microbial species present. The goal of this endeavor is to improve understanding of the interrelationship between geology, geochemistry, and microbiology.

Future research may propose a selective coring program in the lake that will focus on the broad range of features and deposits created by a variety of geologic processes and their details. For example, along the northeastern edge of the lake, landslide deposits are mapped yet the data currently available sheds no or little information to determine whether these deposits represent one landslide event or many. The answer to this question is significant because different scenarios will produce different volumes of water potentially displaced by future landslides of varying volumes. Coring such deposits would help resolve the question of how large a landslide might reasonably be expected, how many events have occurred, when did these events occur, what triggered the landslide events, and when might the next event be. Additionally a detailed modern analysis of cores from Yellowstone Lake would yield detailed insights into the climatic evolution recorded in lake sediments for at least the past 200,000 years. Other important geologic histories that require additional information could include details about hydrothermal domes and explosion craters.

Yellowstone Lake is still a relatively unknown corner of the park. Technology and scientific interest, however, have changed and the scientific foundation established by new mapping will be built upon by future generations interested in the geology and ecosystem of Yellowstone Lake. New developments in ever-higher resolution mapping and imaging, increasingly by autonomous underwater vehicles, will allow more detailed mapping and imaging of smaller features, and monitoring of dynamic crustal movement in the Yellowstone Caldera. New microbiological studies will increasingly link understanding of chemical and biological processes and their effects on the ecosystem. Finally, studies of the effect of potentially toxic elements generated by hydrothermal activity in the lake and elsewhere should be carefully evaluated by studying transference up the food chain, including eagles, osprey, and otters. All of these activities will inform the public and lead to better management of the Park's resources and better understanding of possible hazards under the aegis of the Yellowstone Volcano Observatory.

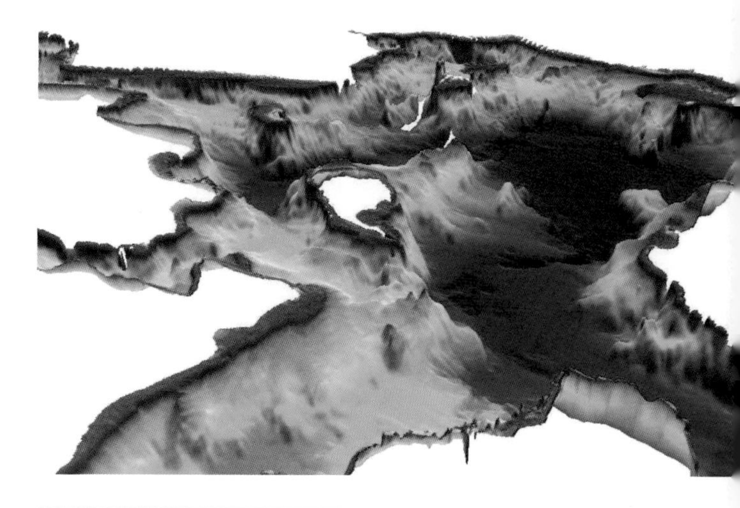

▲ FIGURE 2.4 Several years, multiple scientists, and sophisticated technology helped produce the first comprehensive map of the floor of Yellowstone Lake (view to the NW). (Lisa Morgan, USGS)

ACKNOWLEDGMENTS: The authors acknowledge individuals at the National Park Service (Yellowstone National Park) and the U.S. Geological Survey whose support contributed greatly to the success of mapping Yellowstone Lake. These individuals include John Varley, Mike Finley, Kate Johnson, Ed du Bray, Geoff Plumlee, Pat Leahy, Steve Bohlen, Tom Casadevall, Linda Gundersen, Denny Fenn, Elliott Spiker, Dick Jachowski, Tom Oliff, Paul Doss, and Connie Nutt. Our collaboration with Dave Lovalvo of Eastern Oceanics, Inc. over the past decade has been great and has allowed for unparalleled mapping and exploration of the lake floor. Our partnership with John Varley, Ken Pierce, Bill Stephenson, Sam Johnson, Steve Harlan, Carol Finn, Jim Maki, Pat Bigelow, Jacob Lowenstern, Henry Heasler, Robert Christiansen, Bill Seyfried, Kang Ding, Bill Inskeep, Tim McDermott, and Rich Macur on various aspects of this research has enhanced our understanding of processes associated with Yellowstone Lake. We gratefully acknowledge our collaboration and work with individuals from L-3 Communications, Boris Schulze, Jorg Duhn, Joerg Brockhoff, and Peter Gimpel. We thank Dan Reinhart, Lloyd Kortge, Paul Doss, Rick Fey, John Lonsbury, Ann Deutch, Jeff Alt, Julie Friedman, Brenda Beitler, Charles Ginsburg, Jim Bruckner, Pam Gemery-Hill, Rick Sanzolone, Maurice Chaffey, Dave Hill, Bree Burdick, Erica Thompson, Eric White, Bob Evanoff, Wes Miles, Rick Mossman, Gary Nelson, Christie Hendrix, Tim Morzel, and many others for assistance with field studies. We are grateful to Coleen Chaney, Debi Dale, Maggie Gulley, Joan Luce, Mary Miller, Vicky Stricker, Maggie Gulley, Sandie Williamson, and Robert Valdez for their skillful assistance with project logistics. We are grateful for the constructive reviews by Brad van Gosen, Bruce Heise, and Jerry Johnson. This research was supported by the U.S. Geological Survey, the National Park Service, and the Yellowstone Foundation.

Chapter 3

Using Yellowstone's Past to Understand the Future

Chapter 3

Wildfire is universally understood to be a critical process in most forested ecosystems. Ecologically, it is as important as water, soil, and sun. Today, in the Greater Yellowstone Ecosystem and across the West, policy makers are calling for information about how vegetation and natural disturbances, like fire, will be altered by current and projected climate change. Cathy Whitlock, through her study of ancient pollen, charcoal and other fossils preserved in the sediments of lakes and wetlands, is gaining new insights about climate's role in long-term ecosystem dynamics. Collaborating with ecologists, archeologists, and climate change specialists, she seeks to better understand the role of climate and humans in shaping past plant communities and fire activity. A significant finding of Whitlock's work is that ecosystems respond to climate variations that occur over a wide range of spatial and temporal scales, including year-to-year weather conditions conducive to fire, decadal droughts that cause tree mortality and fuel build up, and long-term climate changes that shape plant communities and fire regimes. Understanding climate change in its many expressions is an important part of planning and adapting to modern and future conditions. Whitlock's work provides useful insights for resource managers as they grapple with the question of how to provide stewardship to our public lands in the face of future climate changes.

J. Johnson

The Runoff at Cutoff Creek (Trey Ratcliff)

Chapter 3

Using Yellowstone's Past to Understand the Future

Cathy Whitlock

Cathy Whitlock, Department of Earth Sciences, Traphagan Hall, Montana State University, Bozeman, MT 59717; Email: whitlock@montana.edu

Information on more of Cathy's work can be found at: http://sites.google.com/site/msupaleoecologylab

Yellowstone's spectacular scenery is the outcome of cataclysmic volcanism, the waxing and waning of large glaciers, and carving the landscape by mighty rivers. The forest and grasslands that blanket the region are a relatively recent addition to the Park, forming about 15,000 years ago on the heels of the last ice age, but they too have been shaped by past geologic events.

Given its dynamic setting, Yellowstone is a wonderful playground for someone who thinks about environmental history, as well as an area of special concern when contemplating the impact of current and future threats. Questions about the natural resilience of the biota to environmental surprises such as wildfire or drought, the rate at which plant communities are capable of forming and dissolving, and the sensitivity of species to changing climate conditions are as relevant for understanding the past as they are for the future. Gaining knowledge about the history of vegetation, fire, and climate is part of the science of paleoecology - the study of past interactions between plants, animals and the physical world. Paleoecologic insights about long-term ecosystem dynamics have been central in the discussions that underlie important management decisions in the Yellowstone region, and they are proving pivotal as we confront current and projected climate changes and other human-induced threats.

This chapter provides an overview of paleoecologic research in the Park over the last three decades. I begin by describing how a paleoecologist "sees" a place like Yellowstone and formulates research questions that motivate field and laboratory work. Some of the significant findings on Yellowstone's history are discussed next, including the role of geology and climate change and the importance of past fires in shaping the ecosystem. Finally, I discuss why knowledge of the past has become an essential part of evaluating the current and future status of the Yellowstone region.

What Do We Do and Why Do We Do It?

Historical sciences, like paleoecology, often answer questions through an iterative testing of multiple working hypotheses. Plausible hypotheses (explanations) are formulated at the outset of the study, and data are used to evaluate the merits of each hypothesis and reject those that don't hold up to scrutiny. Some of the hypotheses are rejected outright, others are modified in light of new discoveries, and new hypotheses emerge during the course of the investigation. In the case of Yellowstone, our research and testable hypotheses have focused on how ancient organisms individuals, populations, and communities responded to changes in the natural ecosystem of the past. The idea is that understanding past associations and interactions provides insight into the present and the future. We are also interested in understanding the hierarchy of climatic and nonclimatic drivers that have shaped the environment while recognizing that these drivers operate at a variety of scales, ranging from continent-wide, slowly-varying changes in climate to local and abrupt disturbances that briefly affect individual watersheds. Another issue of interest has been the role of geology in shaping the history and current distribution of vegetation. Is the dynamic geology of glaciers, earthquakes or geological structure a more important driver of vegetation change, for example, than climate or biological interactions? Finally, does location within the Yellowstone region affect the sensitivity of vegetation to environmental change, and if so, what does this bode for the future?

Natural lakes and wetlands are the best source of information on the history of terrestrial environments. Pollen, charcoal and other fossils preserved in the layers of sediment provide a script of changes that have occurred in the watershed over thousands of years. These layers can be recovered by removing cores from the sediments. The history begins with the deepest layers, deposited when the lake was first formed, and ends with the top layer deposited in the current year. Most of Yellowstone's lakes were created during the final melting of the Yellowstone ice cap, about 15,000 years ago. In the intervening millennia, most lakes have deposited layers of sediment up to eight meters thick. In choosing which lake to sample, we consider its location and how representative the local vegetation, geological substrate, and climate are of a particular region. We are hoping to capture the range of environmental and biotic diversity that exists in Yellowstone.

To study the layers of history, sediment cores are collected, and their fossils and other constituents are examined. A variety of hand-operated coring equipment is available for this purpose, and the selection of an appropriate device depends on the study question and the historical time span needed to answer that question. Information about recent environmental changes, for example, can be found in the upper meter of sediments at most sites. These are recovered in a simple tube fitted with a piston lowered into the sediment. Another version is a meter-high metal box is filled with dry ice. The ice is dropped down the box, the sediments freeze to the outside and samples are recovered. Research concerning the entire 15,000 years of history requires more complicated coring equipment, including a square-rod piston corer attached to metal drive rods. This device takes cores that by five centimeters and one meter long in vertical succession. Each drive is brought to the surface, and the sediments are extruded and wrapped in plastic wrap and aluminum foil before transport to the lab. After each drive, another rod is added to drill the next meter of sediments. Lake sediments are usually the consistency of toothpaste, green or brown in color, and have an earthy odor. Hitting an impenetrable surface usually indicates that bedrock has been reached, and the drilling is finished. About eight people or three packhorses are required to transport the gear to a lake. The actual coring process involves three or four people and in the summer, it takes place from a platform built across two inflatable rafts anchored in the middle of the lake. In winter, we use the ice surface as our coring platform, but deep snow and frigid temperatures often limit access.

▲ PHOTOS 3.1 & 3.2 Cores in the Yellowstone region are often collected in the winter when the crew can work on the frozen lake surface (above). (Melynda Harrison) A hole is be drilled through the ice and the coring device is lowered into the sediments to retrieve a series of sediment cores. Each core is extruded and wrapped in plastic for transport to the lab (below). The cores consist of soft consolidated sediment and the layers describe the environmental conditions at different times in the past. (Jack Fisher).

Although fieldwork may be the most fun, it is the research questions that motivate the project (and usually the funding) and the selection of sites and methods. A good research question is one that has importance beyond the boundaries of the study. Ideally, it is one that addresses a timely or broader scientific question or that proposes to look at old findings with a new, possibly transformative approach. In addition, such a research question must be answerable by a carefully crafted study and thoughtful selection of study sites. A good question ultimately leads to new questions and research directions.

Back in the laboratory is where the hard work takes place. The cores of sediment are unwrapped, sliced

▲ PHOTO 3.3 The sediment cores contain pollen and charcoal that are analyzed in the laboratory to reconstruct the climate, vegetation, and fire history. This core was taken from a floating mat at the edge of Blacktail Pond in northern Yellowstone National Park. (Christy Hendrix)

longitudinally, photographed and described. We then subject them to various types of analysis in order to learn three things about the past ecology - watershed history, age, and the prehistoric ecology of the lake. The lithology or physical composition of the core provides our first clue about the history of the watershed. For example, the base of most records consists of inorganic clay and silt. This indicates rapid deposition in a sparsely vegetated landscape and unstable slopes. Higher up the core, closer to the substrate surface, the sediments become organic, and this transition marks the time that the climate warmed, the lake became biologically more productive, and the watershed was stabilized by soils and plants. Samples are taken to determine the organic and carbonate (CO_3) content of the sediment and the mineral content, all of which can tell us about levels of biological activity in the past. Moreover, variations in the carbonate content of the sediments are often a good measure of chemistry and pH changes related to water temperature.

Paleoecologists are obsessed with time, even though our ability to measure it is often not very precise. Chronologies for our historical information are based on radiometric dating methods, including Carbon-14 (^{14}C) or Lead-210 (^{210}Pb) age determinations. Radiocarbon measurements are made on small wood or leaf fragments of terrestrial plants, on charcoal particles, and sometimes on the organic component of the lake sediments. Yellowstone lakes also contain volcanic ash layers, most notably the ash from the eruption of Mount Mazama, 7676 years ago that led to the formation of Crater Lake in southwestern Oregon. Refinements in radiometric age determinations are continually underway, and through time, the precision and accuracy have improved.

The assemblages of pollen grains in the core are the record of past vegetation. Pollen is produced by angiosperms (flowering plants) and gymnosperms (seed-producing plants). Not surprisingly, wind-pollinated species produce lots of pollen each year and more of it is deposited in the lake sediments than that of insect-pollinated species. Sediment samples are taken at regular intervals in the core, and these are treated with a variety of acids and bases to dissolve all the constituents except the pollen grains. The residue of pollen is mounted on glass slides and examined under the microscope at magnifications of 400-1000x. Different plants produce distinctive and often ornate pollen grains between 25-100 microns in diameter that are identified by comparing them with modern reference material and illustrations in published atlases.

Typically, 300-400 pollen grains are identified and tallied for a given sample in the core. It takes a trained analyst two or three hours to "count" all the pollen grains on a microscope slide. Our ability to identify a pollen grain to a particular plant species or genus is variable and limits our interpretation in some cases. For example, grass

▲ PHOTOS 3.4 & 3.5 Photomicrographs of Douglas-fir pollen (above). (Rudy Nickmann) and charcoal particles (below). (Tom Minckley)

pollen cannot be identified below the taxonomic level of family (Poaceae), so it is not possible to determine whether the grass pollen comes from alpine or steppe species. Pine pollen in the Yellowstone region can be divided into two groupings: lodgepole/ponderosa (*Pinus contorta* or *P. ponderosa*) and whitebark/limber (*Pinus albicaulis* or *P. flexilis*) types. Other taxa, like meadow rue (*Thalictrum*), are secure identified to the level of genus. The presence of seeds, needles and other plant remains in the core often provide species identifications in cases where pollen cannot. In Yellowstone, pine dominates the pollen record because the conifers produce large amounts of pollen in the early summer, and the grains, which have two large air bladders, are easily carried by the wind. Sagebrush (*Artemisia*) also produces abundant pollen, and it blooms in late summer. A typical pollen record will include about 50 different trees, shrubs, and aquatic plants, but most of the pollen grains come from pine and sagebrush.

Pollen counts at each level are converted to percentages and accumulation rates. Changes in the proportion of different taxa through time are the basis for interpreting past vegetation. Because pollen does not have a 1:1 relationship with the plants that produce it, modern studies are needed to interpret past pollen assemblages. Modern pollen information comes from the surface sediments of lakes, and there are hundreds of samples from the U.S. to provide necessary calibration. Modern pollen samples, for example, indicate that pine pollen is overrepresented given the abundance of the tree in the watershed, but Douglas-fir (*Pseudotsuga mensiezii*) pollen is underrepresented.

Fire-history information is based on changes in the abundance of charcoal particles preserved in lakes, and our research in Yellowstone has help guide fire history investigations around the world. For ten years following the 1988 fires, we monitored the input of charcoal into lakes in burned and unburned watershed and were rewarded with a rich understanding of how charcoal data register fire patterns across a large natural region. Information from the 1988 fire has improved our interpretation of longer charcoal records in terms of past fire activity. For example, we can determine how often the area burned, the intensity of the burn, and the fuel type.

We begin by extracting charred pieces of wood and leaves from contiguous one-centimeter intervals of the core and examining them under the microscope. The data are converted to charcoal accumulation rates (number of particles/$cm^{-2} yr^{-1}$). Slowly varying changes in charcoal abundance describe long-term variations in fuel types. For example, forests produce more charcoal than tundra. Charcoal peaks represent individual fire episodes, and they can be summarized to calculate fire-episode frequency. Shifts in abundance between grass and wood charcoal disclose changes in fire regime from surface fires, which burn the ground vegetation, to crown fires that advance through the tree tops with serious consequences to the forest.

My interest focuses primarily on past vegetation and fires, but other fossils and chemical constituents in

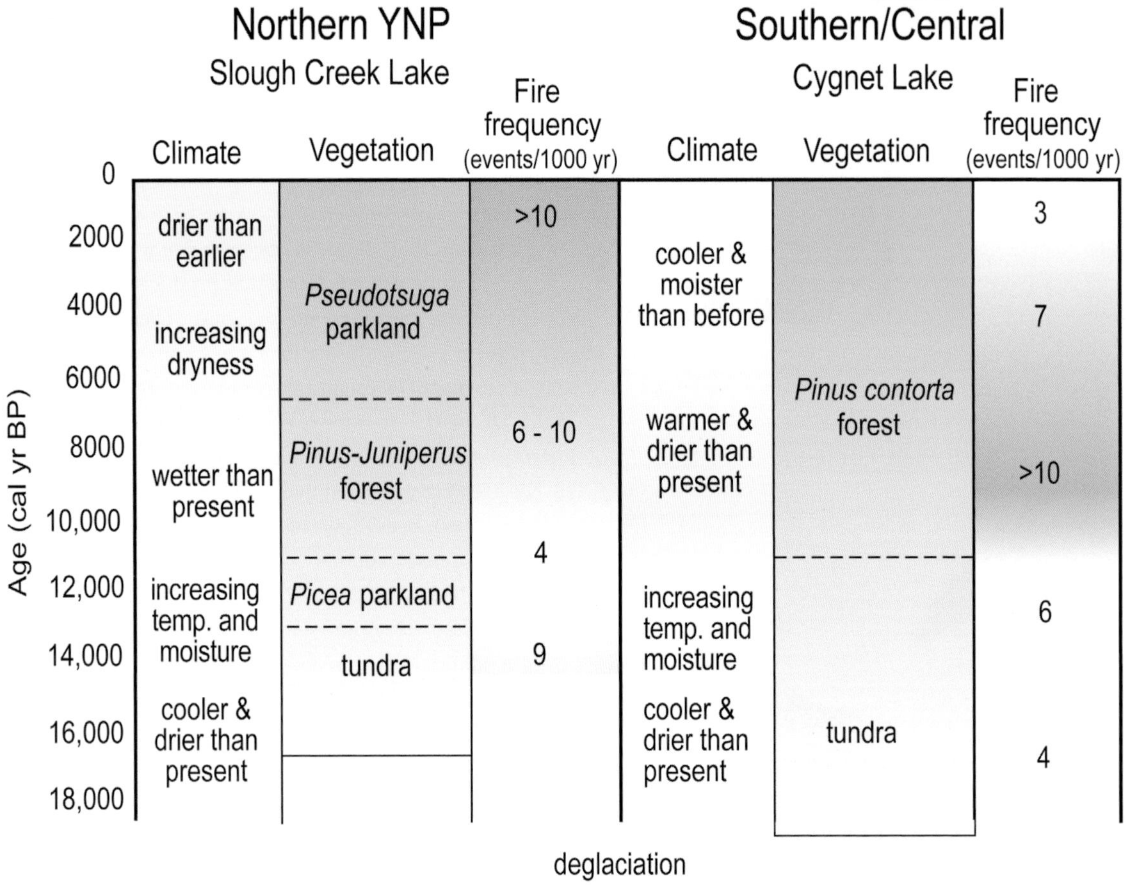

Northern YNP

Slough Creek Lake

Climate | Vegetation | Fire frequency (events/1000 yr)

Southern/Central

Cygnet Lake

Climate | Vegetation | Fire frequency (events/1000 yr)

Age (cal yr BP)

0
2000
4000
6000
8000
10,000
12,000
14,000
16,000
18,000

drier than earlier

increasing dryness

wetter than present

increasing temp. and moisture

cooler & drier than present

Pseudotsuga parkland

Pinus-Juniperus forest

Picea parkland

tundra

>10

6 - 10

4

9

cooler & moister than before

warmer & drier than present

increasing temp. and moisture

cooler & drier than present

Pinus contorta forest

tundra

3

7

>10

6

4

deglaciation

▲ FIGURE 3.1 The climate, vegetation and fire history at Slough Creek Lake near the Lamar Valley and Cygnet Lake near Hayden Valley are based on pollen and charcoal analysis of radiocarbon-dated sediment cores. The pollen data suggest the development of open Douglas-fir parkland in the last 7000 years at Slough Creek Lake as the climate become drier and fire frequency increased. At Cygnet Lake, lodgepole pine forest has been present for the last 11,000 years. Fires were most frequent between 11,000 and 7000 years ago. Fire activity has decreased in the last 7000 years as the climate has become cooler. (Sarah Millspaugh)

lake sediments are examined by specialists to reveal different aspects of ancient environments. Diatoms, the skeletons of microscopic algae, reveal changes in the lake biota that can be tied to changes in nutrients, water temperature, pH, and light penetration. Carbon and oxygen isotopes trace the history of water inputs and evaporation through time that are related to climate. Insects and other animal remains document changes in insect outbreaks and zoological activity, and chemical changes reflect nutrient and erosion inputs related to variations in watershed characteristics. Multidisciplinary studies, like those underway at Crevice Lake in northern Yellowstone, are collaborative efforts to use many data sets to reconstruct the past.

What Have We Learned?

From the first studies in the 1970s to ongoing investigations, we are gaining a better understanding of the Yellowstone ecosystem and its sensitivity to environmental change. New findings about Yellowstone's past range from information about the plants and animals that colonized deglaciated landscapes to insights on how Yellowstone's climate history fits into the larger picture of climate change in the western U.S. For me, both local and large-scale reconstructions are equally rewarding lines of inquiry. It is thrilling to stand on the shores of Crevice Lake and imagine the forest changes over several thousand years. It is also stimulating to join multidisciplinary teams, including paleoclimate modelers, to compare the evidence of past climate change across the continent. What follows are four topics where paleoecologic studies from Yellowstone have led to discoveries that were initially unexpected, but now seen to be scientifically quite important.

GEOLOGY MATTERS

Today, the vegetation of Yellowstone is strongly controlled by geology and its influence on soil conditions. The broad volcanic plateaus of the region support nutrient-poor, well-drained soils, derived from rhyolitic rock. The lack of calcium and potassium and the dryness of the soils limit the success of most conifers on these surfaces. Lodgepole pine over most of its biogeographic range is considered a disturbance-adapted tree, but on rhyolitic substrates (and other well-drained soils), it is the dominant conifer from early to late stages of forest development, generally in the absence of other competitors. Yellowstone's lodgepole pine forests are considered subalpine in the strict sense, and on more nutrient-rich soil types, comparable elevations would support forests of Engelmann spruce *(Picea engelmannii)*, subalpine fir (*Abies bifolia*), and possibly Douglas fir. Another way of thinking about this is: had the massive volcanic eruptions not occurred and laid down broad plateaus of rhyolite, the forests (and landscapes) of central Yellowstone would have resembled those of the nearby Teton and Gallatin ranges. The vast areas of lodgepole pine forest thus are an important legacy of the Yellowstone hot spot and provide a direct feedback to the fire history of the region.

Pollen records from the rhyolitic region shed some light on the importance of geology on the development of central Yellowstone's forests. Following an early period of tundra vegetation, which developed as the last glaciers were melting about 15,000 years ago, the region was invaded by lodgepole pine, and lodgepole pine forests have persisted ever since. This is despite climate changes that have transformed the vegetation in other parts of the Park and region. This attests to lodgepole pine's resilience (some call it a weed) to variations in climate when it occupies poor-quality substrates that limit competitors.

In contrast, pollen records from intermediate andesitic substrates in the south and east, and nutrient-rich calcareous glacial deposits in the north, reveal a more dynamic vegetation history. Following the early tundra period, those regions were colonized by a subalpine forest of spruce, fir, and whitebark pine. As the climate continued to warm, lodgepole pine and Douglas fir moved into the region. We believe that between 11,000 and 7,000 years ago, Yellowstone experienced warmer conditions than today, and this is evidenced by the abundance of Douglas fir pollen at sites that are presently too high for it to grow. As the climate cooled and became wetter during the last 7,000 years, the pollen data suggest that spruce, fir, and pine forests became more common again. If Yellowstone were not a hot spot with active rhyolitic volcanic eruptions, the history of the central region would surely be like the southern sites, and much of the Park would be covered by forests of spruce, fir, and pine. If the region had not been glaciated, the northern part of the Park, in particular, would not have been mantled by glacial till deposits rich in calcium and potassium, and the Northern Range (so-called because this is the northern winter range of Yellowstone elk) would have been a closed forest unsuitable for the winter range of ungulates it supports today. What makes Yellowstone so interesting is that all of these geologic events did occur and combined to shape the present-day ecosystem and ultimately the management challenges of tomorrow.

FIRES HAPPEN

Studies of annual rings of living trees in the Park suggest that the last large fires, comparable to those of 1988, occurred between 1690 and 1730. This is a rather short time span for understanding the frequency of large infrequent fires, and we thought that perhaps the abundance of charcoal particles in lake sediments might allow us to extend the fire history beyond the tree-ring archive. Our modern charcoal studies, initiated with the 1988 fires, helped guide our examination of the ancient charcoal record and how it might be interpreted. In essence, the 1988 fires allowed us to "calibrate" charcoal records by providing some basic methodological information: How charcoal particles got into lake sediments? How far did they travel? What size of charcoal best described a local fire? What processes introduced charcoal into the lake, and how long did it take for burial in the sediments?

A fire-history study of Cygnet Lake, located on the rhyolitic Central Plateau, was undertaken by Sarah

▲ PHOTO 3.6 This photo was taken during a fire in 1939 in the Lewis Lake area. Evidence of this and previous fires is found in the layers of charcoal deposited in lake sediments. (NPS, Yellowstone National Park)

Millspaugh as part of her dissertation. Her research showed that many peaks of charcoal (i.e., fire episodes) occurred in sediments dating between 11,000 and 7000 years ago. In the last 7000 years, the charcoal data suggest that fire frequency steadily decreased leading to the present interval of 300-400 years between large events. The increase and then decease in fires is remarkable, considering that the pollen data indicate that the lodgepole pine forest surrounding the site did not change. This juxtaposition of changing fire occurrence in the absence of changing fuels shows how closely fire activity is tied to climate; fire regimes can change even when forest composition does not. Cygnet Lake and other charcoal-based fire studies in the western U.S. make us realize that large fires do not occur at regular predictable intervals. Instead, the charcoal records indicate that fire frequency has tracked climate change over millennia, and as climate has changed so too has the frequency of fire events. This understanding of long-term fire dynamics is perhaps one of the most important discoveries that came out of the 1988 fire research.

CLIMATE MATTERS

It is not possible to appreciate Yellowstone's ecological history without recognizing the importance of climate changes that have affected the entire western U.S. From independent data and paleoclimate modeling, we know that the presence of continental ice sheets and variations in solar radiation through time have affected latitudinal temperature gradients, shifts in the location of storm tracks, and the strength of large-scale atmospheric circulation patterns and precipitation regimes. One prominent control of climate since the last ice age has been the variations in the seasonal cycle of insolation (incoming solar radiation) caused by variations in the tilt of the Earth's axis and the season of perihelion (when the Earth is closest to the Sun). Between 11,000 and 7000 years ago (the early Holocene), perihelion occurred in summer rather than in winter as it does today, and the tilt of the Earth was greater than present. As a result, insolation was 8.5% greater in summer and 10% less in winter at 11,000 years ago at the latitude of Yellowstone. By 6000 years ago, summer insolation was still higher than present, but less than before, and by 3000 years ago, levels were close to modern. The early-Holocene amplification of the seasonal insolation cycle caused an expansion of the northeastern Pacific subtropical high pressure system, which intensified summer drought in the Pacific Northwest. Conversely, strengthened summer monsoonal circulation in the American Southwest at that time increased summer rainfall in that region.
These large-scale climate changes are evident in the vegetation and fire history of Yellowstone.

The early Holocene is the most extreme warm period of the last 15,000 years, and the climate impact on the Yellowstone region has received a lot of research attention. Today, Yellowstone lies at the transition between areas under the influence of the subtropical high and those strongly affected by summer monsoons. As a result, there are two precipitation regimes in the Park at

▲ FIGURE 3.2 This landsat satellite image shows that large fires, like those of 1988, are characterized by a mosaic of different burn severities. The 1988 fires left some forest stands untouched while others were destroyed. Postfire vegetation recovery in the past can be studied from pollen analysis. (USGS, Earth Resources Observation and Science Center)

present. The southern and central region and the highest elevations have a Pacific Northwest climate (so-called summer-dry region), wherein the influence of subtropical high leads to relatively dry summers. The northern part of the Park receives more of its annual precipitation in summer as a result of its comparatively drier winters and greater summer monsoonal activity (summer-wet region), and summer thunderstorms are also more likely in this region.

Paleoecological data and paleoclimate modeling indicate that the two precipitation regimes were strengthened in the early Holocene by the higher-than-present summer insolation. The southern and central parts were drier than today, because the subtropical high pressure system was stronger then than it is now. Northern Yellowstone was effectively wetter in summer as a result of the intensified monsoons. Charcoal records from sites in the northern part of Yellowstone show low fire activity then, progressively more fires in the last 7,000 years as the region dried. Sites in the central and southern Park, including Cygnet Lake, indicate highest fire activity in the early Holocene when the aridity was greatest, and fewer fires since then as cooler wetter conditions established. Sites in the northern range, like Slough Creek Lake, show the opposite pattern of fewer fires before 7000 years ago and more fires since then.

The ideas developed in Yellowstone have become an organizing framework for paleoclimatic research in the western U.S. Follow-up studies, testing the summer-wet/summer-dry climate hypotheses, have been conducted in the Wind River Range, the Bitterroot-Selway, the southern Rockies, and the Beaverhead Range. Together, these studies provide a much clearer picture of how the climate system works and the influence of mountains in creating regional differences.

Why is the study of climate conditions 15,000 years ago important? It suggests that adjacent regions can have quite different climate responses as a result of the local influence of continental-scale climate changes. The spatial patterns result from the interplay between atmospheric circulation and mountainous landscapes. Similar complex interactions will surely be an important factor with future climate change. Evidence for the persistence of two precipitation regimes offers testable research hypotheses about the availability of resources for wildlife and prehistoric peoples at different times in the past. Such hypotheses can help direct interdisciplinary research at a broader scale. Consider the Yellowstone Lake region in the early Holocene as an example. Humans were living in and around Yellowstone during the early Holocene but the archeological evidence is too sparse to reconstruct their activities. We expect that their resource base was strongly dictated by the climate and environment. Given our understanding of the paleocological record, early-Holocene winters were colder than today as a result of low winter insolation, and perhaps as important, mid-winter thaws were infrequent. Such conditions would have discouraged overwintering of game in the Park, and winter food sources for year-round occupation would have been scarce. A rapid rise in spring and summer insolation in the early Holocene would have led to an early thaw, rapid snowmelt, and early ice breakup on Yellowstone Lake. Spawning of Yellowstone cutthroat trout (*Oncorhynchus clarki bouvieri*) in the tributary streams is tied to the timing of peak stream flow and this would have occurred earlier in the year in the early Holocene.

In contrast to the severe winters, summer and fall conditions had many attributes that would have encouraged human occupation. Frequent fires in the summer-dry regions of Yellowstone would have kept much of the forest in early successional stages, although average fire size may have been relatively small. Based on forest development following modern fires, the diversity of birds and mammals in young forests would have generally been high, which may have improved hunting success in the summer. Yellowstone Lake temperatures would have been higher-than-present in summer and fall, and fish may have been an important early summer food source (although there is no archeological evidence of this to date). Dry warm weather in fall would have extended seasonal occupation. Greater radiative fog from a warm lake, however, may have shifted the location of campsites away from the shore to rocky promontories during the fall. So, one can hypothesize that a warmer climate, a warmer lake, and more open forests would have enhanced food resources between 11,000 and 7000 years ago compared to present.

HUMANS MATTER (BUT HOW?)

We've been involved in two investigations to ascertain whether changes in land management were significant enough to be registered in the recent environmental records of Yellowstone. The first study was done eighteen years ago, when colleagues from the University of Minnesota and I undertook a study of the last 200 years of Yellowstone's history. We were interested if there was evidence of environmental change in the sediments of small lakes in the northern range following the creation of national park in 1872. In particular, we were looking for possible effects caused by changes in winter ungulate populations, such as increased erosion or nutrient enrichment in lakes during times of high elk and bison numbers. If such evidence was found, they might be attributable to changes in Park management,

▲ PHOTO 3.7 Climate change in the region threatens the health of whitebark pine - an important source of food for many animal species including red squirrels, Clark's nutcrackers, and grizzly bears. Across its range, nonnative blister rust and native mountain pine beetle have attacked whitebark pine. Increased temperatures projected for the future are likely to increase whitebark pine mortality in the Yellowstone region. (Big Sky Institute, MSU)

particularly to natural regulation, which began in earnest in the 1970s. The research involved sampling the uppermost sediments of five small lakes in the Northern Range, dating the sediments with ^{210}Pb dating methods (which is good for the last 200 years), and analyzing the fossils and chemistry of the sediments for signs of environmental change. The largest signal that we obtained came from Floating Island Lake near Tower Junction. There, detection of erosion came only at the time of road construction. At other sites, the diatom assemblages indicated times of nutrient enrichment that might have been caused by heavy ungulate use at the lakes, but the timing of enrichment was not synchronous from one watershed to the next. In the case of aspen, some records showed a decrease in aspen pollen in the 20th century, but others indicated no change or even a slight increase. None of the records detected a decline in aspen in the last 150 years, and we surmised that aspen is poorly represented in recent centuries in comparison to its prominence at the end of the glacial period. Some lakes revealed changes in sedimentation rate, but again, no widespread change was detected that might correspond with historic fluctuations in elk number.

This study has been more recently expanded upon by the investigation at Crevice Lake in the Yellowstone River canyon. Pollen, geochemistry, paleomagnetism, and diatoms records, examined at decadal resolution over the last 2600 years, suggest that the last 200 years of environmental history has been relatively complacent, compared to dramatic adjustments in the lake and watershed that occurred in response to climate changes between BC 150 and AD 1100. We can find comfort from the combined findings of two investigations, which imply that Yellowstone today is a relatively pristine, naturally functioning ecosystem. Human-related impacts of the last two centuries have left little trace in the sediments of lakes, and by all measures, most pale in comparison to natural variations of the more distant past.

Yellowstone now faces threats from an ever greater human footprint in the region and globally, and projected changes in the region seem likely to exceed the natural range of variability. Future climate model projections scaled down to the Yellowstone region suggest changes in temperature and precipitation that will affect the distribution of Yellowstone's species, assuming that they can keep pace. Present climate projections focus on the consequences of doubling CO_2 in the atmosphere - something that is now projected to occur sometime this century. Under a $2xCO_2$ scenario, model projections suggest that the Yellowstone region gets warmer in both summer and winter, summers get drier (as a result of warmer temperatures), and winters become wetter as levels of greenhouse gases increase. Warmer, wetter winters imply more rain, less snow, earlier snowmelt, and longer summer conditions. Measurements taken in the last 20 years suggest that this climate trend is already underway.

The biogeographic range shifts associated with future climate change are dramatic and will not be easily attained given the fragmented landscape of the region. Species' responses will also be complicated by sharp elevational and latitudinal gradients and trade-offs in precipitation discussed previously. High-elevation organisms will likely be the most impacted because climate warming will shift suitable habitat to ever higher elevations, and eventually, no elevation in the Yellowstone region will be high enough to sustain viable populations. Whitebark pine is an example of a highly vulnerable species in this regard, and its viability is further compromised by outbreaks of mountain pine beetle (*Dendroctonus ponderosae*). The seeds of this conifer are an important food resource for grizzly bear (*Ursus arctos*), red squirrel (*Tamiasciurus hudsonicus*), and its loss could lead to the collapse of a vital ecosystem in the Park.

Future changes in the biogeographic range of tree species raise equally great concerns for Park resource managers. For example, some species are projected to find suitable conditions in the Yellowstone region in the future that don't grow there today. For example, warmer conditions may allow Gambel's oak (*Quercus gambeli*) to extend its range northward from Utah and Idaho into northwestern Wyoming. At the same time, projected warmer wetter winters resemble those found today in the interior Pacific Northwest, and habitats suitable for wet-loving conifers, like western larch (*Larix occidentalis*) could also exist in Yellowstone in the future. Thus, the studies to date suggest that some species will survive with little change (e.g., lodgepole pine), the ranges of others will shift from the south or north, and some species (e.g., whitebark pine) will be lost from the region altogether.

Such equilibrium-based climate and species projections are unsatisfying on several levels, and there is an urgent need to improve them. For example, it is not clear if species are capable of naturally adjusting their range at the rate required to keep pace with projected climate change. Studies elsewhere suggest that species will have to move 40 times faster than anything we've seen in the fossil record. Movement today and in the future will also have to occur across diverse landscapes with many barriers and different land uses. What will be the impact of fragmentation? And, if the rates are too rapid, what opportunities does this create for non-native species and weeds to invade weakened native communities? Human intervention and directed assistance are used on commodity lands to move economically important species to new areas, but is this appropriate for national parks and wilderness areas as well? Certainly, the answers to these questions depend to some degree on particular land ownership and management goals, but they also point to the need for discussion among different stakeholders, coordinated responses, and stepped-up monitoring and inventory efforts. We are at the stage in resource management where climate change science will become a part of every natural resource policy discussion in the Greater Yellowstone Ecosystem and other natural ecosystems.

Conclusion

Paleoecology is a fascinating subject in its own right, but, more than that, the scientific discoveries that it provides are the foundation for understanding current and future ecosystem dynamics. Without information about the natural range of environmental variability, it would be impossible to evaluate changes evident today or likely to occur in the future. It would also be difficult to assess human impacts, including the consequences of management decisions at the local scale or anthropogenic climate change at the global scale. In the Yellowstone region, knowledge of the past has led to renewed appreciation of the importance of geology, climate and climate change, and natural disturbance in shaping the diversity of plant and animal communities that exist today. It has also pointed to new research areas to examine how prehistoric peoples and early Euro-Americans may have utilized natural resources in the past. Finally, paleoecologic information provides insights into the range and rate of environmental changes that will help us evaluate current human-induced change.

Not all is known about Yellowstone's past, and in most respects, our understanding is rudimentary and incomplete. For example, recent drought in the region is profoundly altering the landscape and drying up wetlands for the first time in Park history. What is not clear is whether drying of this magnitude has ever occurred before and what the lasting ecological consequences might be. Such information would help resource managers to assess the current peril. We also have limited knowledge about the processes that link different components of the ecosystem. What, for example, are the natural drivers of limnologic change? Is it changes in terrestrial inputs, changes in climate, or biotic interactions within the lake itself? And, what are the linkages between climate change, fire activity, and the infestation of forest-insect pathogens, such as occurring at present.

Our ability to address nuanced questions about ecosystem processes is aided by new techniques, better conceptual and numerical models, and well-dated high-resolution data sets. The paleoecologic research that began in the Yellowstone region in the 1970s still holds the key to critically important questions, and those mysteries will motivate creative scientific investigations for decades to come.

▲ PHOTO 3.8 Winter recreation, such as skiing and snowmobiling in the Greater Yellowstone Ecosystem, are likely to be affected by climate change as the region experiences reduced snowpack and fewer cold days. Some businesses may be able to diversify the types of recreational activities they offer based upon the changing climate. Communities where recreation is an important component in the local economy and lifestyle may face significant challenges in the near future. (Jerry Johnson)

Chapter 4

Understanding Grizzlies: Science of the Interagency Grizzly Bear Study Team

Chapter 4

Chuck Schwartz is a very lucky man. Over the course of three decades he has worked with grizzly bears in Alaska, Russia, Pakistan, and Japan and he has never had a serious accident. Good technique helps. Chuck is the leader of the USGS Northern Rocky Mountain Science Center Interagency Grizzly Bear Study team based in Bozeman, Montana. His group conducts the long-term research and monitoring of grizzly bears in the Greater Yellowstone Ecosystem. Before coming to Montana, he worked for the Alaska Department of Fish and Game for more than 20 years conducting research and providing management recommendations on moose, brown, and black bears. As a member of the IGBST, Mark Haroldson has studied the bears of Yellowstone for over 30 years; probably no one knows the population as well as he does. Kerry Gunther is Yellowstone's bear management biologist. When human-bear conflicts occur, Kerry is the guy who deals with it. He has worked in the park for over 25 years.

The history of the region is bound up with the grizzly. Frank and John Craighead began a long-term research program of the bear in 1959 and developed techniques to capture and track large animals. They pioneered the use of radio telemetry and early satellite imagery data.

Between 400 and 600 grizzlies are spread across the Greater Yellowstone Ecosystem. Their range and low density makes them a particularly difficult animal to study. Grizzly bear demographics, particularly the role of human caused mortality, is central to the mission of the IGBST. The most important demographic group is females with cubs; they add to the stock of bears and therefore are critical to the recovery and sustainably of the population. The passion Chuck holds for the bears is obvious in his writing and in a career dedicated to knowing "the Great Bear".

J. Johnson

Yellowstone Grizzly Bear (Steve Hinch)

Chapter 4

Understanding Grizzlies: Science of the Interagency Grizzly Bear Study Team

Charles C. Schwartz, Mark A. Haroldson, and Kerry A. Gunther

Charles C. Schwartz, U.S. Geological Survey, Northern Rocky Mountain Science Center, Interagency Grizzly Bear Study Team, Forestry Sciences Lab, Montana State University, Bozeman, MT 59717; 406-994-5043; FAX 406-994-6416; Email: chuck_schwartz@usgs.gov

Mark A. Haroldson, U.S. Geological Survey, Northern Rocky Mountain Science Center, Interagency Grizzly Bear Study Team, Forestry Sciences Lab, Montana State University, Bozeman, MT 59717; 406-994-5042; FAX 406-994-6416; Email: mark_haroldson@usgs.gov

Kerry A. Gunther, Bear Management Office, Yellowstone National Park, WY 82190; 307-344-2162; FAX 307-344-2211; Email: kerry_gunther@nps.gov

The official web site of the IGBST is: http://nrmsc.usgs.gov/research/igbst-home.htm

The grizzly bear** (ursus arctos horribilis) **inspires fear, awe, and respect in humans to a degree unmatched by any other North American wild mammal. Along with other bear species, it has the capability to inflict serious injury and death to humans and sometimes does.

In the Greater Yellowstone Ecosystem, some grizzly bears live in areas visited by crowds of people or near human settlements. Here, the presence of the grizzly remains a physical and emotional reality. A hike in the wilderness that includes grizzly bears is different from a stroll in a forest from which grizzly bears have been purged; nighttime conversations around the campfire and dreams in the tent reflect the presence of the great bear. Contributing to the aura of the grizzly bear is the mixture of myth and reality about their ferocity, unpredictable dispositions, large size, strength, huge canines, long claws, keen senses, swiftness, and playfulness. Bears share characteristics with humans. They have a generalist life history strategy like extended periods of maternal care and, and omnivorous diets. They are highly intelligent, learn quickly, and have long memories.

The dominance of the grizzly in human imagination played a significant role in the demise of the species. Conquest of the western wilderness seemed synonymous with destruction of the great bear. Prior to European settlement of North America, grizzly bears could be found from northern Alaska south through Canada and the western United States and into northern Mexico. In the contiguous United States, habitat was altered or destroyed and important bear foods like salmon, elk, and bison were greatly reduced by dam building, market hunting, and competition with livestock. Primarily during the 1920s and 1930s, the grizzlies' historical range decreased nearly 98%. Of the 37 grizzly bear populations known to exist in 1922, 31 were gone by 1975. In the West, grizzly bears were poisoned, shot, and trapped to reduce depredation on domestic cattle, sheep, and poultry. A stockman captured the prevailing attitude in the 1920s: "The destruction of these grizzlies is absolutely necessary before the stock business...could be maintained on a profitable basis.".

Yellowstone National Park (YNP) was established in 1872 to protect the area's geysers, thermal features, and scenic wonders. However, due to its remoteness and the protections afforded by national park status, it also became one of the last refuges for grizzlies in lower forty-eight states. Grizzly and black bears became one of the parks most popular attractions. By the 1880s park visitors enjoyed watching bears that gathered to feed on garbage dumped behind the hotels. As early as 1907, park staff were killing some black and grizzly bears because of conflicts with people. By 1910, black bears had learned to panhandle for food from tourists traveling in horse-pulled wagons. The first recorded bear-caused fatality occurred in 1916, when a grizzly bear killed a wagon teamster in a roadside camp.

When cars replaced horses and wagons, the number of park visitors and the amount of garbage they left behind increased. More garbage attracted more bears and park managers even encouraged bear viewing at some dumps by providing log bleachers and interpretive rangers. Unfortunately, this mix of people interacting with food-conditioned bears created problems. From 1931 through 1969, bears caused an annual average of 48 human injuries and 138 incidents of property damage. After a bear killed a woman in the Old Faithful Campground in 1942, Congress criticized park managers for failing to solve the bear problems.

In 1963, an Advisory Committee to the National Park Service issued a report titled "Wildlife Management in the National Parks" that recommended maintaining park biotic communities in as near a primitive state as practical. It recommended and nearly complete removal of human influence on wildlife populations to allow natural processes to work. The Leopold Report, in combination with the fatal mauling of two women by grizzly bears in separate incidents in Glacier National Park, the frequency of bear-caused injuries and property damages in YNP, and new environmental regulations for open-pit garbage dumps, led to the implementation of an intensive Bear Management Program in YNP in 1970. In addition to strict enforcement of regulations prohibiting the feeding of bears, the new program called for bear-proof garbage cans and dumpsters and the closure of all the park's garbage dumps.

In 1970, the decision to close the park's last two garbage dumps was highly controversial. Brothers John and Frank Craighead, pioneers of grizzly bear research, agreed that the dumps were inconsistent with National Park Service management philosophy, but believed they played a crucial role in reducing human-caused bear mortality. They opposed a rapid phase-out of the dumps, especially the Trout Creek Dump. They believed an immediate closure of all dumps would increase conflicts, management removals, and mortality both inside and outside the park. The Park Service believed a gradual phasing out of dumps would result in several more generations of bears becoming dependent on human foods, leading to more bear-conflicts over time. After obtaining the advice of the National Sciences Advisory Committee, park authorities chose to close the park's remaining two dumps quickly in 1970 and 1971. The state of Montana closed the three dumps in the park gateway communities of West Yellowstone, Gardiner, and Cooke City in 1970, 1978, and 1979, respectively.

Within twelve years (1968–1979), all municipal dumps in the GYE that had aggregations of grizzly bears were closed, and many bears that previously ate garbage dispersed in search of alternative foods. Many of the bears that came into conflict with people at developed sites, campgrounds, private homes, and on cattle and sheep allotments were removed by the National Park Service or the state fish and game agencies from Wyoming, Montana, and Idaho, or were killed by private citizens. At least 140 grizzly bear deaths were attributed to human causes during 1968–71. Due to the growing disagreement between the Craighead brothers and the park over the dump closures, and restrictions placed on their research and publications that the brothers did not accept, their research permit in Yellowstone was not renewed after 1971.

▲ PHOTO 4.1 The grizzly bear is one of the most adaptable large mammals in North America. Extremely resourceful, the bear will feed on almost any source of calories including carrion, insects, vegetation or, human garbage. This bear is likely digging for ground squirrels in the Lamar Valley. (NPS, Yellowstone National Park)

Due, in part, to uncertainty about the status of Yellowstone bears and declines in other grizzly bear populations, the U.S. Fish and Wildlife Service listed grizzly bears in the lower forty-eight states as a threatened species under the Endangered Species Act in 1975. Once a species is listed under the ESA, the U.S. Fish and Wildlife Service is required to prepare a recovery plan and lay out a framework for recovery. The foundation of this framework is built on a scientific understanding of the species, its habitat requirements, limiting factors, and population trend.

Creation of the Interagency Grizzly Bear Study Team

The need for better information was motivation for the creation of the Interagency Grizzly Bear Study Team (IGBST) in 1973. The Study Team initially had representatives from the National Park Service, the Forest Service, and the U.S. Fish and Wildlife Service; representatives from the states of Wyoming, Montana, and Idaho were added later. Dr. Richard Knight was named the Study Team leader by Assistant Secretary of the Interior Nathaniel Reed. The primary objectives of the team are to determine the status and trend of the grizzly

▲ PHOTO 4.2 Bears eating garbage at Trout Creek, Yellowstone National Park, ca 1960s. Early in the Park's history, feeding garbage to bears was common. This practice was stopped in the late 1960s. (NPS, Yellowstone National Park)

bear population, the use of habitats by bears, and the relationship of land management activities to the welfare of the bear population.

For more than 30 years, members of the Interagency Grizzly Bear Study Team have been investigating grizzly bear biology in the Greater Yellowstone Ecosystem. Much of the early work was gleaned by tracking radio-collared bears, examining scats and foraging sites, and observing bears in general. In recent years, the Study Team has used the newest research techniques and cooperated with outside specialists in chemistry, genetics, and nutrition to further advance the understanding of grizzly bear ecology. The new research techniques used by the Study Team include highly-accurate Global Positioning Satellite (GPS) collars that pinpoint a bear's location many times a day; hair snares fashioned of barbed-wire that collect small samples of hair when bears rub against them. DNA and nutritional analyses that determine the sex, identity, and the diet of each can be preformed on very small samples, such as bone flakes, a drop of dried blood, or a few hairs from the hair trap.

Biology of the Yellowstone Grizzly

Grizzly bears are a difficult animal to study. They have often been relegated to remote rugged terrain, occur at low densities, are wide ranging, and elusive. Grizzly bears have one of the slowest rates of reproduction of any large terrestrial mammal in North America. In the GYE, female bear typically do not reach sexual maturity until 4–7 (5.8 years) years of age, and produce a litter of 1-3 cubs once every 3 years. They are long lived with reproductive senescence occurring around 28 years of age. Young bears dependent upon their mothers care have low survival rates with only 57% reaching age two. Cause of death for most dependent young is often unknown, but most probably die from starvation or are preyed upon by other, often male, bears. About 85% of adult bear mortality in the region is human caused. Agency removal of bears that have come into conflict with people either by euthanasia or relocating to zoos is the major cause of mortality (54.2%). Other causes include self-defense kills

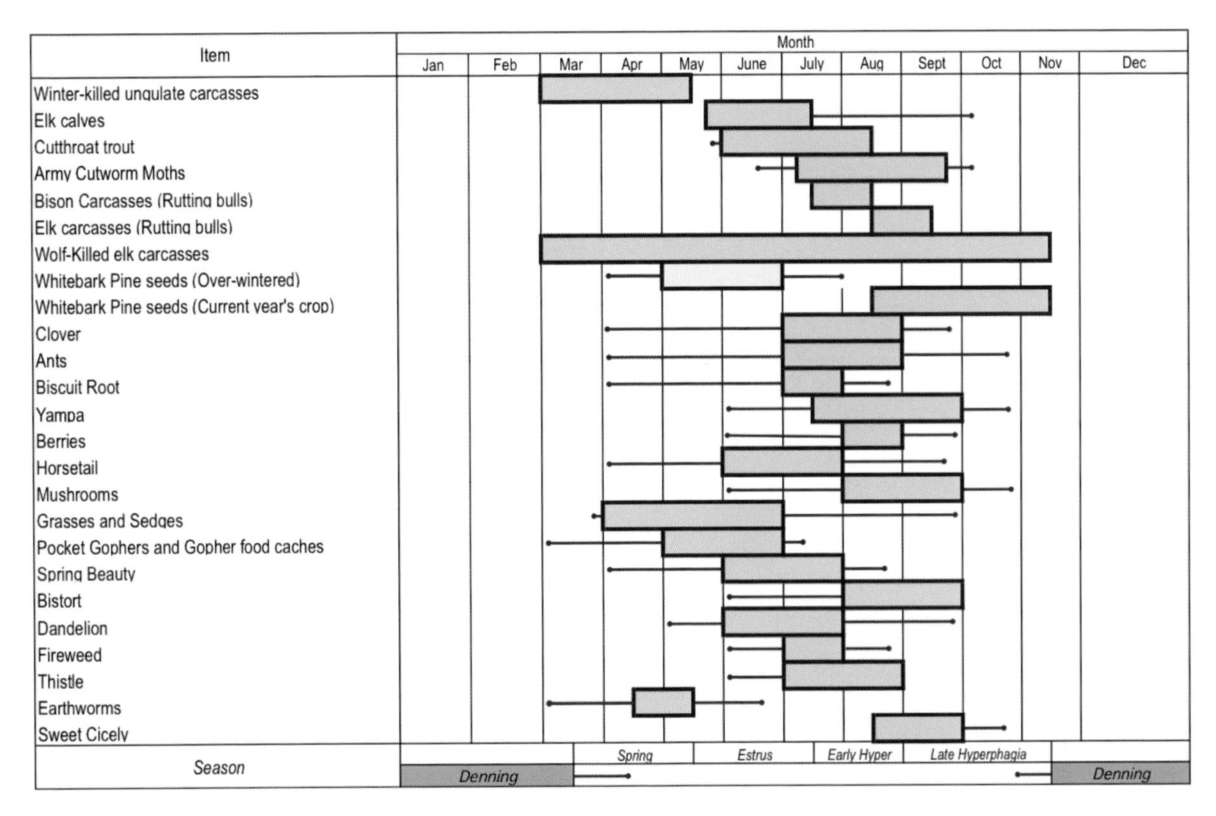

▲ FIGURE 4.1 Seasonal availability of grizzly bear foods in the Greater Yellowstone Ecosystem. Yellowstone grizzly bears spend 5–6 months in a winter den. When not denning, bears spend much of their time foraging on a large number of different foods. Bears make use of these foods when they are available. (IGBC, USGS)

by big game hunters (17.0%), mistaken identity kills by black bear hunters, (8.5%), malicious killing or poaching (18.6%), and road kills (1.7%).

Grizzly bears are truly seasonal animals. They have evolved life history strategies that included denning and associated physiology in response to adverse environment conditions, primarily seasonal lack of food and unfavorable weather. In Yellowstone they spend 5–6 months each year in winter dens; they typically entering dens in late fall and do not emerge until early spring. During hibernation, they do not eat, drink, urinate or defecate. They subsist entirely on fat stores from the previous autumn. Pregnant females are the first to enter dens, typically during the last week of October or the first week of November. Females with young (cubs or yearlings) enter dens in early November, and male bears den about a week later. In spring, males are the first to emerge, typically during the last week of March, although it is not uncommon for some males to emerge as early as February. Females with older cubs in the den emerge around the third week of April, whereas females with newly born cubs typically do not leave their den until the last week of April, and some remain in their dens through April.

Grizzly bears are opportunistic omnivores. They deviate from most other meat-eating carnivores by the volume and variety of vegetative foods in their diet. Yellowstone grizzly bears commonly consume herbaceous vegetation during spring and early summer. In early spring, bears seek out carcasses of both bison and elk that die of starvation during the long winter. Ungulate carcasses are a highly valuable food for bears in the spring. As the season progresses bears take advantage of other high quality foods as they become available. Yellowstone grizzly bear actively prey on newly born elk in late May through June. They also consume other animals when available. Ground squirrels, pocket gophers, and ants are relatively common items in the diet. During summer, their diet is a mix of plant and animal matter.

In late summer, grizzly bears enter the hyperphagic period - from the Greek for "excessive eating". Because bears spend nearly half of the year hibernating, they must store large quantities of fat to sustain physiological functions during denning. Females that give birth to cubs in the den require additional fat for milk production. In years when fall foods are abundant, it is not unusual for bears to enter their dens with 35-40% of their body mass as stored fat. To build their fat stores, Yellowstone grizzlies need to eat large quantities of high quality foods: these include seeds of whitebark pine, army cutworm moths, and ungulates.

Research priorities under the ESA are driven by the mandate to recover and sustain the population of bears in the region. Since listing in 1975, the emergent agenda includes monitoring the bear population, tracking and understanding their health, the nature of their food supply, breeding and genetic issues, and tracking individual bears in order to better understand their behavior and habitat use. Research scientists have devised a number of techniques to track and study grizzlies. Techniques range from simple observations, to sophisticated use of radio telemetry, visual analysis of bear scats for food habits to more complicated chemical processes. Here we describe just a few.

▲ PHOTO 4.3 Adult females with cubs add to the stock of grizzly bears in the region and are the focus of much of the research of the Interagency Grizzly Bear Study Team. The number of cubs a female produces varies based on the availability of food and other environmental conditions. (NPS, Yellowstone National Park)

Females with Cubs

When the Yellowstone grizzly was listed as a threatened species, biologist recognized that to recover the population, human caused mortality had to be minimized. At the time, reducing adult female mortality by one or two bears per year would likely have been enough to stabilize the population. But managing mortality within sustainable levels required knowledge of how many bears resided within the ecosystem. However, estimating bear numbers in not an easy task.

For the first two years after its formation, the IGBST was not permitted to capture and/or mark bears in YNP. This early prohibition against marking individuals eventually led the Study Team to develop a method for estimating population size that the team continues to use today. Dr. Knight and the Study Team observed that adult females with cubs were easy to see and that the number of cubs provided clues for distinguishing family groups. Summing the count of unique females over three successive years provided a conservative estimate of how many reproductive (i.e., adult) females were in the population. Counts were added over three years because, on average, adult female grizzlies produce a litter every three years. This produces the best available abundance data in the GYE.

To distinguish unique females from repeated sightings of the same female, the Study Team developed a rule set for observations. It was recognized that these rules were not perfect and if errors occurred, two different females were more likely called the same female as opposed to calling two sightings of the same female two different families. Thus, it was felt that employing the rule set returned conservative (or low) estimates for the number of females. This method was adopted as part of the Grizzly Bear Recovery Plan in 1993. A running three year average of females with cubs was used to establish a minimum population number and set allowable mortality limits. However, using counts of unique females with cubs was criticized by some scientists because the rules to differentiate families had not been verified, the technique did not account for variation in observer effort (number of people looking for females) or the sightability of bears with area and time (bears tend to be more easily seen in dry years), and the estimate was a minimum count not an estimate of the total population.

During the late 1990s, the Study Team and numerous collaborators began investigating methods to address these concerns. An evaluation of the rule set used to differentiate families confirmed that the method resulted in increasingly conservative estimates as population size increased. Methods to estimate total numbers of females with cubs and account for variation in sightability of bears and observer efforts were also investigated. Employing the best of these methods indicates an estimated 5% increase in grizzly females with cubs during 1983–2006. The Study Team working with other professionals also devised method to take estimates of females with cubs and derive an estimate of total population size. These estimates are now used to set mortality limits for cubs and yearlings (dependent bears), and independent females and males.

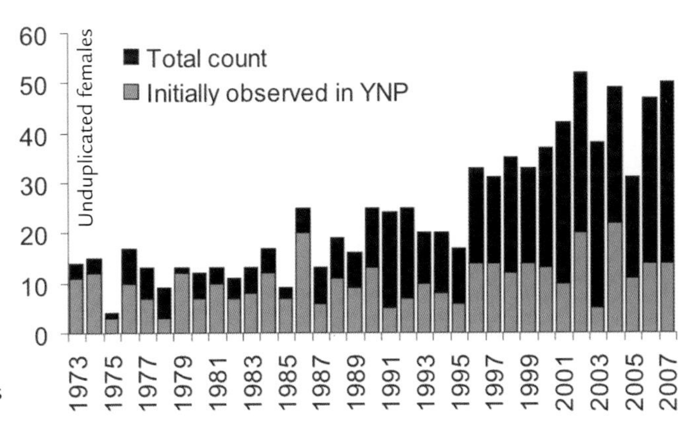

▲ FIGURE 4.2 Changes in numbers of females with cubs-of-the-year in the Greater Yellowstone Ecosystem (total count) and within Yellowstone National Park, 1973–2007. Changes in counts of females with cubs-of-the-year reflect the general increase in the Yellowstone grizzly bear population for about 1983 forward. The light colored portion of the bars indicates a stable population of bears in YNP. The darker portion indicates and expansion of the population to the Greater Yellowstone region. (IGBC, USGS)

Estimating Vital Rates from Radio-Marked Bears

Probably the most widely employed technique to study wildlife is radio telemetry. In 1959, brothers John and Frank Craighead and their dedicated team developed many of the earlier methods used to safely capture, immobilize, age, and mark grizzly bears. Nearly 50 years ago, they developed the first radio-transmitter collar and directional receiver and tracked two grizzlies to their winter dens.

Biologists use telemetry systems to track animals. Telemetry, derived from the roots *tele* = remote, and *metron* = to measure, is composed of three parts: the radio transmitter, an antenna, and radio receiver. The transmitter is attached to the animal with a collar and transmits a signal at a very high frequency (VHF). Each collar has its own frequency similar to different channels on a conventional radio. By tuning the receiver to the specific channel, the signal transmitted from the collar can be heard. With the use of a directional antenna, instrumented animals can be located either from the ground or with the aid of an aircraft. Once an animal is found, many variables can be recorded. These include, the number of cubs or older offspring seen with a collared female (this allows us to estimate reproductive rates and cub and yearling survival), the habitat type the animal is using, and the current status of the individual (i.e., alive or dead). Our radio collars are built with a mortality mode. This mode allows the researcher to determine if the animal is alive or possibly dead. The system is simple. If the animal moves the collar at least once every four hours, the transmitter returns a signal (beat) every second. If however, the animal is dead or has lost its collar, the beat rate declines to once every two seconds. Tracking to the collar allows us to determine if the bear is dead or if it has shed its transmitter. Collars typically transmit for about 3 years, but many are shed before the battery powering the transmitter fails. We intentionally incorporate a cotton spacer in all the collars. This cotton spacer typically rots through in about two to three years and the collar falls off. This allows us to recover the collar and have it refurbished so it can be used multiple times. It also means the bear does not wear the collar for the rest of its life.

The Study Team began capturing and radio-collaring grizzly bears in 1975. Early efforts were limited because of the time and expense required to capture, instrument, and monitor the bears. Aircraft were required to locate and monitor the status (i.e., alive or dead) of collared bears and to obtain observations of females for estimates of reproductive performance. In 1986 the Study Team began collaring bears specifically for the purpose of monitoring population trend. The initial target was to monitor 10 adult females that were well-distributed throughout the ecosystem. However, because of their larger home ranges, male bears were captured about four times as often as females, providing additional information on topics including habitat use, movements, and cause of mortality. But it is female bears that drive the demographic vigor of the population.

In the mid-1990s, the target was raised to 25 monitored females to allow more precise estimates and increase confidence in the results. By then, estimates of adult female survival and population trend suggested that the population had stabilized but disagreement persisted over whether the population was likely increasing. An analysis published in 1999 that used reproductive data and survival rates obtained from 1975–1995 suggested the population had changed little to none during that period. Subsequent work published by the Study Team and collaborators clearly demonstrates that GYE grizzly bear numbers increased at an average annual rate of about 4–7% during 1983–2001. This increase is likely a result of increased female survival and is similar to trend estimates derived from counts of females with cubs. The agreement between these two methods that used independent approaches provides confidence that the increase in the population was real.

Determining What Is Hazardous to Bears

By following radio-instrumented bears the Study Team has been able to understand what influences bear survival. Location, it turns out, is very important to bear survival. We have constructed survival models that reveal bears living inside Yellowstone National Park have very high rates of survival, whereas bears living near human developments have lower rates of survival. But why? By looking at variations in bear survival and linking this to measures of habitat quality and disturbances within each bears home range, we are able to show that bear mortality is more influenced by human disturbances on the landscape. Thus as the number of roads, recreational developments, and homes within a bears home range increase, their chances of mortality also increase. By combining this hazard surface with our estimates of reproduction and survival, we created what biologists refer to as a source-sink surface. This surface considers the rate of survival necessary to maintain a healthy population. Source habitats are where lambda ≥ 1, whereas sink habitats are those where lambda < 1. These models allow managers to evaluate and sometimes mitigate impacts of changes to bear habitat for the overall health of the bear population.

GPS Telemetry — Knowing Where Your Bear Is at Midnight

Nearly 50 years ago, Frank and John Craighead used the very first radio-transmitter collar and directional receiver to track two Yellowstone National Park grizzly bears to their winter dens. Since that time, we have witnessed numerous improvements, innovations, and technological advancements in animal tracking systems. Scientists realized early on that since it was possible to determine a satellite's orbit from earth, it was quite feasible to reverse the process to determine an exact location on Earth. This concept spawned the creation of a satellite-based navigation and positioning system. The Global Positioning System (GPS) technology offers positional accuracy better than 30 meters.

In 1995, the first GPS collars were deployed on Alaskan brown bears. GPS collars were deployed on Yellowstone grizzly bears two years later. GPS telemetry has revolutionized grizzly bear research, improving our ability to collect abundant, accurate, fine-scaled spatial data. VHF telemetry required a person to physically locate the bear, so most locations were collected during the day when conditions were suitable for aircraft flights. This meant that most bears wearing VHF collars were located between the hours of 6:00 AM and 2:00 PM. We knew very little about their nighttime activity and habitat use patterns. GPS technology is capable of collecting a position fix at any time and thus allowed for data

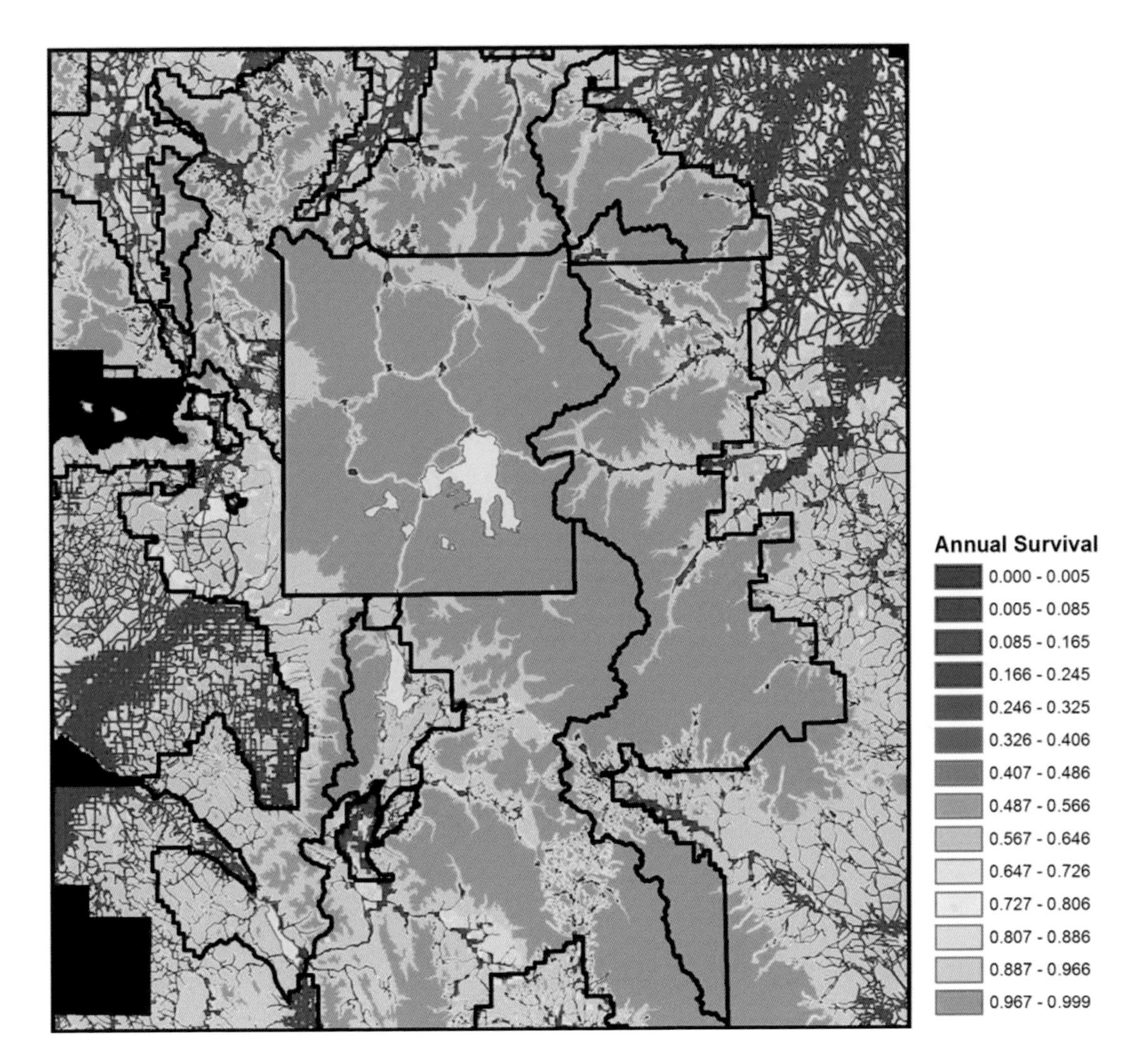

▲ FIGURE 4.3 The survival of a grizzly bear is directly related to where it lives in the Greater Yellowstone Ecosystem. Using data from radio-collared bears, the Study Team was able to determine what factors influence survival. Bears living in areas with human development (roads, homes, campgrounds, lodges etc) and in areas open to autumn ungulate hunting have lower rates of survival when compared to bears living in more secure habitats (wilderness areas and national parks). This figure shows predicted rates of survival for an adult female in the GYE. Warmer colors show areas of high predicted survival, whereas cooler colors show areas where bear survival is compromised by human development. (IGBC, USGS)

collection 24 hours a day, 7 days a week. This wealth of information has revealed some interesting behavioral patterns in the Yellowstone grizzly. For example, using GPS technology and incorporating an activity switch in the collar, the Study Team has been able to reconstruct the daily activity patterns of individual bears. These patterns show that grizzly bears are most active at sunrise and sunset, but are also active during mid-day and at night. Interestingly, male bears tend to be more night active. Activity patterns are also linked to daily air temperatures. Grizzly bears appear to be heat sensitive and reduce their daytime activity patterns when air temperatures exceed about 20°C.

DNA Fingerprinting and Mercury Analysis

Cutthroat trout were previously an important food for grizzly bears living around Yellowstone Lake, but cutthroat numbers have declined precipitously since the illegal introduction of lake trout there. Counts of spawning cutthroat trout at Clear Creek declined from

Estimating the Population Trend

Scientists estimate population change with some fairly complicated mathematical equations. Our sample of radio-collared bears allows us to estimate vital (estimates of reproduction and survival) rates. These rates include an estimate of reproduction success for female bears (fecundity). We derive this estimate by observing and tracking the fate of cubs born to collared females. Collared bears also provide estimates of annual survival. From the sample of marked individuals and the application of survival models, it is possible to determine the average survival rate for different classes of bears. We typically estimate survival for males and females, and compare these rates for adult and subadult bears. Once we have estimates of vital rates, these are combined in mathematical models such as a Leslie Matrix or Lotka equation to estimate population trend.

A simple analogy may make this more understandable. We can think about the grizzly bear population in Yellowstone as a bank account. The population represents the amount of money in this account. Reproduction in the population is the same as interest paid on the principal. New money deposited increases and withdrawals reduce the size of the account. Estimating population change is simply tracking new bears entering the population (reproduction) and bears leaving (mortality). The best expression of trend for a population is Lambda (λ) or "finite rate of change". Estimates of λ tell us whether, on average, numbers of births and recruitments for a population are greater than deaths or visa versa. Thus, $\lambda > 1$ indicates an increasing population, $\lambda = 1$ stable, and $\lambda < 1$ a decreasing population. A population that remains stable (neither grows nor declines), has a trajectory of 1.0. This would be equivalent to a bank account where withdrawals equal the interest paid to the account. A declining population has a trajectory of less than 1.0. A population with an estimated trajectory of 0.9 is declining at 10% per year; we've withdrawn the interest paid to the account plus 10% of the principal. However, population change is much more sensitive to the loss of an adult female than the loss of a cub. Adult females produce cubs and thereby add to the "capital stock" of bears, whereas a cub must remain in the population for at least five years before it can begin to produce offspring. If we put this into dollar terms, the loss of an adult female is equivalent to withdrawing 73¢ whereas the loss of a cub is only about 13¢, or the loss of one adult female has the same potential impact on the population as the loss of five cubs. It's like getting interest paid on the account each year or waiting five years before any is paid. Obviously, the account with annual interest grows faster. Biologists estimate reproductive and mortality rates from radio-collared animals and can determine population trajectory, just like you do when you check your bank account statements.

more than 70,000 in 1978 to around 500 in 2007. Studies of fish use by bears in the late 1980s relied on detecting fish parts or determining the presence of fish remains in bear scats. In the late 1990s, the Study Team discovered that mercury in the effluent from thermal vents in Yellowstone Lake could be used as an indicator of fish consumption by bears. When a bear eats a fish that has eaten plankton containing mercury, the mercury is deposited in its hair. By working with scientists from the Washington State University Bear Research, Education, and Conservation Program in Pullman, WA we were able to determine that measuring the concentration of mercury in bear hair provides a direct measure of the number of fish consumed by that bear. Coupling mercury concentrations in bear hair with DNA analyses has allowed biologists to estimate how many

▲ PHOTO 4.4 Researchers work quickly and according to strict standards of care when their research requires them to drug and handle a bear. They often administer oxygen and IV fluids. Here, scientists monitor oxygen saturation, heart rate, and temperature as they carry out isotope and bioelectrical impedence tests on a young bear. (IGBC, USGS)

bears consume fish, how many fish each bear eats, and the sex of the bears that eat fish. Results showed that in the late 1990s most fish were eaten by male bears. A three-year study, started in 2007, is documenting the extent to which bears have shifted away from fish to other foods. Preliminary results confirm that very few bears still eat fish, and that most of the bears that previously ate fish are now focused on preying on elk calves adjacent to the lake. Elk are now calving in the post-fire blow-down resulting from the 1988 fires and tracking studies suggest that the bears have shifted accordingly.

Stable Isotopes and Bioelectrical Impedence

For years, biologist learned about the food habits of the Yellowstone grizzly bear by examining scat and visiting telemetry locations of bears in an attempt to determine what they were eating. Within the past decade, new methods are now available to supplement this information. This work took on greater importance with the effects of climate change on regional vegetation and the discovery of lake trout in Yellowstone Lake in the early 1990's.

The new technique we use to quantify diets of both living and dead bears is called "stable isotope analyses." Isotopes are different forms of the same element, for example ^{14}N and ^{15}N. Both are nitrogen but the far rarer form, ^{15}N, has one extra neutron, is non-radioactive, and occurs naturally. When an animal eats plant material that contains small amounts of ^{15}N and digest it, the body preferentially retains ^{15}N relative to ^{14}N. Thus, bears that have eaten only plants will have ^{15}N in their hair or bones similar to levels found in deer and elk. However, when a bear eats meat (e.g. elk), it is consuming a food with higher levels of ^{15}N because the herbivore flesh contains a higher concentration of ^{15}N when compared to plants. Consequently, bears that eat meat have ^{15}N levels elevated above those found in herbivores. It is this ^{14}N to ^{15}N ratio that allows us to quantify the proportion of plant and animal matter that a bear ate during the past few weeks, months, or lifetime. By feeding the captive bears at Washington State University various diets that included deer, trout, clover, grass and other foods and analyzing the isotope ratios of both food and bear, we were able to calibrate this technique specifically for grizzly bears.

The Bear Research, Education, and Conservation Program at Washington State investigated the historical diets of Yellowstone grizzly bears. The oldest grizzly bear bones they analyzed came from a 1,000-year-old packrat midden excavated from the Lamar Cave. Due to the efforts of this hard-working packrat that had a fetish for bones, they showed that meat (everything from ants to trout and elk) provided 32% of the nourishment for those grizzly bears and 68% came from plants.

From 1914 to 1918 when many hotels were feeding kitchen scraps to attract grizzly bears for tourist entertainment and towns had open-pit garbage dumps, nourishment of the Park's grizzly bears switched almost entirely to meat (85% meat: 15% plants). After all such feeding was stopped by the early 1970's and bears were forced to return to natural foods, the diets of young bears of both sexes and adult females returned to the levels observed 1000 years ago (~40% meat: 60% plants), although adult males have continued a more carnivorous life (~80% meat: 20% plants). Large males can more efficiently prey on the Park's elk and bison or can claim the carcasses of animals that died from other causes. Bears that have been killed for preying on livestock

Using DNA to Estimate Bear Numbers

We are all aware of the powers of DNA fingerprinting. The O.J. Simpson case is a good example. Biologists also use DNA to identify individual bears and their sex. We use barbed wire hair snares to collect samples. The method is quite simple. Bears are attracted to a specific site with bait, usually cattle blood. The blood, referred to as a "call lure" is suspended above the ground in a plastic milk jug from a tree several feet above the ground. It is placed high enough so that a bear cannot obtain a food reward, but is attracted to the site by the smell of the blood carried on the wind. Surrounding the call lure, a corral is built with a single strand of barbed wire set about 2 feet off the ground. When the bear crosses under or over the wire, a small amount of hair is snagged on the barb. Hair with follicles contains bear DNA. Using genetic techniques, it's possible to identify individual bears and determine their sex. Using capture-mark-recapture models (these models estimate the number of bears not captured based on the frequency of bears that are captured), we can estimate the number of bears using an area.

outside the Park had diets that were 85% meat:15% plants. These levels of meat consumption are in contrast to the grizzly bears in Montana's Glacier National Park or Alaska's Denali National Park where plant matter provides 97% of their nourishment. Thus, for grizzly bears, the opportunity to consume meat differentiates the Yellowstone ecosystem from many other interior ecosystems where bears must feed primarily on plants.

One of the most important plant foods eaten by grizzly bears in the Greater Yellowstone are the high-fat, energy rich nuts of the Whitebark pine. In years following a good crop of seeds of this high elevation conifer, grizzly

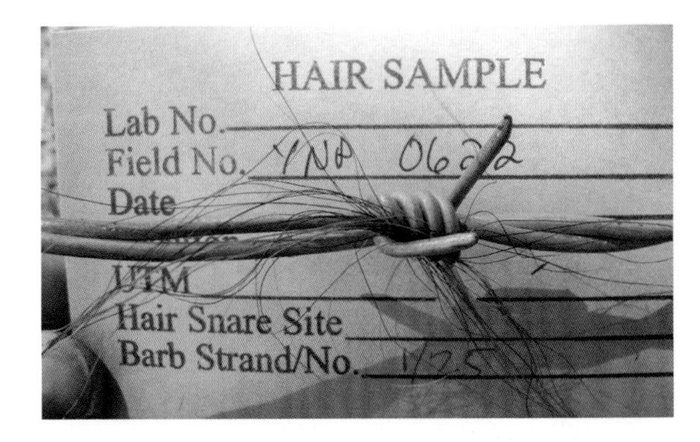

▲ PHOTO 4.5 The hair corral trap is an extraordinarily simple method for collecting DNA from large predators. A barbed wire fence encircles a lure of blood. When the bear investigates the potential food, he leaves a few clumps of hair as he ducks under the wire. The hair is collected and returned to the lab for analysis. The bear is unharmed and avoids the danger of being trapped, drugged and handled. (Justin Teisberg)

bear females tend to produce more three-cub litters than one-cub litters. The opposite is true following poor seed crops. In poor seed years, bears in YNP shift their diets, and their survival rate remains high because the park is a secure environment. However, in years of poor seed production outside the park, particularly on the edge of the ecosystem, more bear conflicts occur as they expand their feeding range closer to humans, and mortality rates tend to increase. In a separate study, we wanted to actually quantify the nutritional value of pine nuts to grizzly bears. Like the other studies, we needed to find some element that occurred in pine nuts that did not occur in the bears' other foods, was absorbed when nuts were consumed, and ultimately was deposited in the bears' hair in proportion to amount of nuts consumed. Fortunately, whitebark pines concentrate a rare sulfur isotope (^{34}S) that occurs in the nuts' protein and therefore is absorbed by the bears and is deposited in their hair. Using isotope analysis similar to what we employ with ^{15}N we were able to demonstrate that ^{34}S was a good biomarker for quantifying pine nut consumption rates in grizzly bears. This study showed that during the year when cone production was high (average 39 cone/tree) pine nuts provided 97% of the annual nourishment for the Park's grizzly bears. The breakpoint for good versus poor years was about 20 cones/tree. We also showed that when nuts were scarce, grizzly bears ate more meat.

Whitebark pine is currently under attack by native mountain pine beetles, previous outbreaks of which have resulted in high mortality rates in trees across the West. The Study Team, in cooperation with the National Park Service's Inventory and Monitoring Program, is tracking both mortality rates in the GYE due to pine beetles and blister rust infection, an exotic fungus that has killed many whitebark pine trees in the Pacific Northwest since it arrived in North America in the late 1920s. It has been less lethal in Yellowstone, but continues to spread and surveys suggest that about 20% of the whitebark trees in the GYE are infected with rust. We do not yet have statistically rigorous estimates for whitebark pine mortality rates from either blister rust or mountain pine beetles or for the extent of their impacts on whitebark communities for the entire GYE. However, the impact on some whitebark stands from pine beetles appears to be considerable in portions of the GYE. How the changes in whitebark abundance will affect grizzly bear numbers is not entirely known.

In an effort to understand how the decline in whitebark might affect grizzly bears we employ a method called bioelectrical impedance analysis (BIA). BIA is a common method used to estimate body composition. Electrical impedance or opposition to the flow of an electrical current can be used to estimate the amount of water within the bear. We know from other research that there is an inverse relationship between body water and body fat. Using simple equations, we can estimate the amount of fat in a bear. We are interested in knowing how fat a bear is because in the GYE they can spend up to 6 months in a winter den, living entirely off stored body fat. Bears must gain sufficient weight to survive this long denning period, and for females that produce cubs, fat also provides the energy necessary to produce milk during lactation. Our early results suggest that the bears are able to attain adequate fat levels for denning in both good and poor seed years.

Conclusion

In April 2007, the U.S. Fish and Wildlife Service officially removed the grizzly bear in the GYE from the Endangered Species list. As expected, several lawsuits were filed

▲ PHOTO 4.6 The recovery of the grizzly bear is one of the great success stories of conservation. The bear is a living symbol of wilderness and our national commitment to the preservation of our last wild places. (IGBC, USGS)

challenging this decision. Proponents for delisting point to the successes that have occurred since 1975, including the increase in bear numbers, the recolonization of previously occupied habitats, high rates of female survival, and the current health of the population. Those opposed to delisting express concerns about the possible effects of climate change and declines in whitebark pine, and whether delisting the Yellowstone population separately from the other U.S. populations was appropriate. The agencies involved in the process prepared numerous documents detailing how the bears will be managed, including monitoring protocols, mortality limits, and habitat management programs. The courts will now determine if all these efforts meet the requirements of the ESA. Regardless of that decision, the IGBST will continue to monitor grizzly bears in an effort to understand how the species adapts in a dynamic ecosystem in the face of natural and man-made change. The long-term survival of grizzlies in Yellowstone is intimately linked with humans, how we impact the ecosystem and how much space we leave for bears. The challenge of the 21st Century is attempting to avoid and subsequently attempting to correct the errors of the 19th and 20th Centuries. If we are not up to the task, there will be no true wilderness to inspire the thoughts and dreams of the children in the next century, only forests.

Chapter 5

Interactions Between Wolves and Elk in the Yellowstone Ecosystem

Chapter 5

Yellowstone is sometimes called "the Serengeti of North America". The savannah grassland of eastern Africa and Yellowstone are the two best places find large concentrations of herd animals and all the attendant ecological linkages that come with them. For years tourists watched bears, elk, bison, and other large animals hunt, graze, and raise their young. But something was missing. The experience was not quite as natural as it seemed. Elk herds of several hundred individuals could be easily watched in the Lamar Valley during broad daylight; they appeared docile and mostly unconcerned. Outside the park, hunting was relatively easy and success rates for late season cow elk hunts were high and elk numbers remained at record levels. In 1995 and 1996 everything changed when 31 grey wolves were reintroduced to their former hunting ground.

Like any predator, wolves have an impact on prey populations. The question is to what extent. Scott Creel has studied the response of prey animals from predators like the African wild dog, leopards, and lions in Tanzania and Kenya. Scott has studied the relationship between elk and wolves in Greater Yellowstone for almost two decades. Scott's science is a blend of long hours of fieldwork - frequently in winter, physiology, and high tech laboratory analysis of stress hormones.

The grey wolf, long the most despised predator in North America, had an immediate effect. Elk behavior changed quickly - they became wary of anything moving through the sagebrush meadows. Herds became smaller, more mobile; and harder to find during hunting season. Over the next decade assertions were rampant that wolves were killing off the great elk herds of the region and drastic control measures were necessary. Hunters called for eradication. Wyoming declared the wolf a game species and implemented plans for hunts. Pro-wolf advocates pointed to drought, hard winters, and overpopulation as reasons for the decline in elk numbers and many opposed delisting the wolf from the Endangered Species Act. Scott's conclusions about the impact of wolves on the Yellowstone elk is much more complex and nuanced than the political debate would suggest. Hopefully, his study will encourage rational public discourse and result in policies that preserve this magnificent animal.

J. Johnson

Cow Elk (John Winnie)

Chapter 5

Interactions Between Wolves and Elk in the Yellowstone Ecosystem

Scott Creel

Scott Creel, Department of Ecology, Lewis Hall, Montana State University, Bozeman, MT 59717;

Email: screel@montana.edu

Review more of Scott's work at: http://www.montana.edu/wwwbi/staff/creel/creel.html#Creel%27s%20Homepage

Yellowstone National Park is unique in the lower 48 states of America for two main reasons. First is its geology, which has always been unusual. Second, it holds the entire suite of large carnivores that were present at the end of the Pleistocene (11,000 years ago) together with healthy populations of large prey such as elk and bison.

Yellowstone was not originally unusual in this regard, but increasing numbers of people and their increasing levels of consumption have rapidly and effectively squeezed wildlife off of most of the American landscape. Worldwide, when human populations expand, wildlife populations inevitably contract. For two main reasons, one ecological and one sociological, large carnivores like the wolf or the African wild dog are typically among the first to go. Ecologically, large carnivores require large areas with intact prey populations, which in turn require suitable habitat. The range of a wolf pack can easily be several hundred square kilometers, and a viable population requires many packs.

Sociologically, large carnivores often face active persecution, due to potential conflicts with people and their livestock. Despite its status as the world's first legally gazetted national park, Yellowstone has not been immune to the second process. As part of a national program of eradication, the last known wolves in Yellowstone were killed in 1924, and with them, the wolf was effectively extinct in the US portion of the Rocky Mountains. Following an absence of seven decades, 14 wolves in three packs were released inside Yellowstone in 1995, together with a release of 15 wolves in central Idaho. The following year, 16 more wolves were released in Yellowstone. Wolf recovery, and the slower but overlapping recovery of Grizzly bears over recent decades, has restored the Greater Yellowstone Ecosystem

▲ PHOTO 5.1 The reintroduction of the Grey Wolf to the Yellowstone ecosystem has been remarkably successful. Between 400 to 450 wolves currently occupy the Greater Yellowstone Ecosystem. This mature wolf may tip the scales at 40 Kg and live as long as 8-10 years. (U.S. Fish and Wildlife Service).

to a condition that is rather unusual for ecosystems in developed nations, with ecologically functional populations of all of the extant large carnivores native to the region at the end of the last ice age.

In addition to its intrinsic value to Yellowstone wilderness, recolonization of the wolf creates an unusual opportunity to understand the function of a terrestrial ecosystem subject to 'top-down' effects that are initiated at the apex of the food web. This statement should not be read to imply that wildlife populations in Yellowstone are regulated solely by natural processes outside of the activities of humans; humans have strong effects on the ecological processes of Yellowstone (as with virtually all modern ecosystems) and the policy of allowing 'natural regulation' of wildlife populations within the park was only adopted in recent decades - not long in ecological terms. Additionally, processes outside the park inevitably have influences inside the park. For example, most elk migrate out of the park and are exposed to harvest during the hunting season. Bison are trapped and killed in large numbers, with more than half of the population slaughtered in the winter of 2007 – 2008 (as part of a poorly conceived response to the presence of *Brucella* in several species, including the far more abundant and wide-ranging elk). Humans remain the most common cause of death for wolves and bears, but my focus here is on the strong and complex interactions between wolves and elk, where each has dominant limiting effects on the other.

Wolf and Elk Recovery in Yellowstone

As with most reintroductions of large carnivores, wolf reintroduction was controversial. Attention was focused on evaluation of its probable consequences. Since the reintroduction, considerable effort has gone into research to directly measure ecological responses. Prior to the reintroduction, three concerns were commonly expressed. First was the fear of attacks on humans. No such attacks have occurred, which is not surprising, given the long record of human-wolf interactions elsewhere. For example, there are more than 3,900 wolves in Minnesota, Wisconsin and the upper peninsula of Michigan, which is more than triple the size of the Rocky Mountain population (as of December 2006), and these wolves have occupied the Great lakes region for decades with no attacks on humans.

Second was concern about predation on livestock, particularly sheep and cows. As expected, wolf packs that establish ranges outside of wilderness areas have come into conflict with animal agriculture. Predation on sheep is patchy, but local losses can be substantial, particularly when sheep graze high-elevation pastures on public land, with little human presence to dissuade wolves from occupying the area. Predation on cattle is also patchy, but is most common in low elevation grasslands in river valleys where elk congregate in winter. If a wolf pack occupies such an area in the winter, it is likely to produce pups shortly before elk migrate to high elevation summer ranges, leaving the wolves with a local prey base that is suddenly dominated by cattle. Largely as a result of situations like this, human-caused mortality takes more than one-fifth of Montana's wolf population each year, mainly through predator-control operations in response to predation on livestock. Despite these genuine conflicts, wolf predation on livestock in the region has remained low relative to other causes of death (<1% of all livestock losses in the northern Rockies, according to the U.S. Fish and Wildlife Service), and the wolf population has grown numerically and expanded geographically.

The third concern, which has proved well founded, focused on the potential impact of wolf predation on elk populations. Although the causes are very different, Rocky Mountain elk and wolf populations have followed similar trajectories over the last 200 years, both driven by humans. While people were intentionally eradicating wolves and other predators, they were simultaneously (though unintentionally) eliminating elk through over-hunting. Where the journals of Lewis and Clark described herds of thousands, elk had dwindled to only seven relict populations in the entire state of Montana by the turn of the 20th century. Beginning in the early 1900s, programs to reintroduce elk into their former range and to promote population growth allowed elk to recover in the Yellowstone area and elsewhere. By the turn of the 21st century, elk were widely distributed in mountainous areas, and had attained high densities in many places, including the Greater Yellowstone Ecosystem (GYE).

Like almost everything in the conservation and management of large animals in the United States, the return of the elk was attended by complications and controversies. While most hunters favored policies that maintained large numbers of elk, others argued that the Yellowstone population had grown so dense that it was altering the plant community on which it depended. Others noted that by feeding hay to elk in winter, elk populations were kept artificially large, potentially exacerbating conflict with ranching. This is a major catch-22, because the current intention of winter feeding (in the Wyoming portion of the GYE) is not to increase elk numbers, but to keep elk from aggregating on ranches and competing with cattle for food. Finally, in response to clumped food sources, elk cluster at atypically high densities on feed-grounds, thereby creating conditions that may promote the transmission and persistence of brucellosis within the elk herd. *Brucella* infection can induce abortion in cattle (particularly in first-time breeders), so the persistence of *Brucella* in Yellowstone wildlife has become an economic issue, necessitating testing and vaccination programs for cattle. Although the elk population was originally infected by cattle, subsequent vaccination programs have eradicated *Brucella* from the US outside the Yellowstone region.

As these complex, intertwined issues illustrate, there is no widespread consensus on the number of elk that is desirable for the Yellowstone ecosystem; the desired outcome depends on the value that different individuals place on different things. Of course, from the perspective of ecosystem function, adjectives like 'desirable' do not have to enter the analysis. An ecosystem shifts among states through time in a manner that depends on the interactions of the species that are present. There may be equilibrium points to which the system tends to return, provided that driving forces like the climate remain relatively constant, but the modern view is that ecosystems are dynamic, and their state depends on multiple factors, from predator-prey ratios to wildfire and drought.

▲ PHOTO 5.2 This elk calf was born in the Gallatin Canyon portion of the Greater Yellowstone Ecosystem. The presence of wolves leads to broad changes in elk grouping, foraging behavior, habitat selection, diet, and nutrition. These "risk effects" are associated with a decrease in the progesterone levels of female elk during gestation and therefore reduced calf production. (David Christianson)

From the perspective of ecology, the dominant question with respect to wolf reintroduction was, and still is: "How will the addition of wolves alter the elk population, and what consequences will responses by the elk have for other species?" Ecological experiments like a large-scale wolf reintroduction are rare, so it is interesting to go back and consider what answers were offered to this question prior to the reintroduction, how those answers compare to the outcomes that have been observed thus far, and what we have learned.

Wolf Recovery

In terms of rapidly establishing an ecologically functional predator population in a large area, wolf reintroduction in the GYE has been a success. From the release of 31 individuals in 1995 and 1996, the GYE population grew to 376 individuals in 31 breeding packs by December 2006. Geographically, the population expanded to include portions of Wyoming, Montana and Idaho. The Northern Rockies population held 1,243 known wolves and 90 breeding packs at the end of 2006, which constituted 24% of the 5,251 known wolves in the lower 48 states. The Mexican Gray Wolf population in Arizona and New Mexico held only 59 individuals, and the remaining 3,949 wolves were in the Great Lakes population. In addition, Alaska held an estimated 6,000 to 7,000 wolves. In one decade, the recovery of wolves in Central Idaho and GYE has had a substantial effect on national wolf numbers, and particularly on their geographic distribution.

Predicted Responses of Elk Numbers to Wolf Recovery

Prior to wolf reintroduction, there was not complete unanimity about the likely effect on Yellowstone elk numbers, but the most widely-accepted prediction (from the National Park Service's environmental impact statement) was based on the well-studied Northern Range herd, and predicted a decline of 5% - 30%. At the time of reintroduction, the National Park Service summarized this consensus view as follows:

Gray wolves are being restored, but not because park managers think the wolves will "control" the number of elk. Instead, fifteen North American wolf experts predict that 100 wolves in Yellowstone would reduce the elk by less than 20%, 10 years after reintroduction. Computer modeling of population dynamics on the northern winter range predicts that 75 wolves would kill 1,000 elk per winter, but that the elk would be able to maintain their populations under this level of predation, and with only a slight decline in the level of hunter harvest.

In the 1970s, other authors had also argued that predation by wolves would be largely compensatory, meaning that wolves would kill elk that would have died anyway, or that the rates of survival and reproduction of the survivors would improve due to reduced competition for food. In these authors' view, wolves were not predicted to reduce the elk population appreciably. At the other extreme, several authors noted that wolf predation has stronger effects on the dynamics of their dominant prey in ecosystems with multiple predators (such as grizzly bears and mountain lions), and predicted that elk numbers would decline by as much as 50%.

Actual Responses of Elk Numbers to Wolf Recovery

By the winter of 2007 - 2008, elk numbers had declined farther than predicted by any of these studies. The northern range herd has steadily declined from approximately 17,000 in 1995 to less than 7,000 in 2006, a reduction of 60%, or triple the consensus prediction of a 20% decline. Elk in the small, nonmigratory Madison-Firehole population in the center of the park have declined by more than 60%. Elk in the Gallatin Canyon have also declined significantly since 1995, from around 1738 to around 1101 - a 37% reduction. These patterns are striking, but when evaluating population trends, one must keep in mind that no species is limited exclusively by a single factor, and elk are not limited only by wolves. Consequently, one must consider the possibility that the strength of other limiting factors might have increased during the same period. This question has seen considerable attention, and some authors have suggested that predation by wolves is not likely to be the primary cause of the decline, which they attribute to a combination of dry weather and 'supercompensatory' effects of human harvest, in which each elk harvested causes the population to decline by more than one individual.

However, most data suggest that wolf predation is the dominant ecological process driving the decline of Yellowstone elk. First, there is the observation that elk constitute approximately 90% of the prey taken by GYE wolves, and wolves account for more than 90% of the observed predation on adult elk. Grizzly bears also take a substantial number of newborn elk, but bear predation is rare for elk older than a few months. Second there is the abrupt nature of the decline, it's timing, and its relation to trends in elk populations outside of the wolf recovery area. For several decades prior to 1995, elk numbers were rising in the GYE, as in the rest of Montana. Elk populations in areas of Montana with little wolf presence have mostly continued their growth, and many Elk Management Units in Montana are now well above their target population sizes. Overall, there are now more elk in Montana than at any time since the late 1800s. This pattern contrasts sharply with population trends for GYE elk, and strongly suggests that general climatic trends have been favorable for elk in the years since 1998.

Considered mechanistically, it is not surprising that the span of dry years coinciding with wolf recovery has been climatically favorable for elk. A great deal of research shows that winter starvation is a strong limiting factor for elk, and that the strength of this effect is dependent on the severity of winter snowfall. Yellowstone elk feed primarily by grazing on grasses, rather than browsing woody vegetation. Yellowstone elk lose body mass steadily during the winter, and this negative energy balance is exacerbated by long winters with deep or heavy snow. In contrast, recent variation in levels of summer rainfall does not appear to cause enough variation in the amount of grass available to have much effect on elk numbers. Overall, the benefits of low-snow winters have been stronger than the costs of low-rain summers. As an aside, it is notable that most climate models predict increased precipitation for the Northern Rockies, but less snow accumulation due to warmer temperatures. If this pattern does emerge, one would expect elk populations to increase. Whether these changes will be enough to offset the effects of wolf predation will be an interesting research question for the future.

Many Yellowstone elk migrate out of the park to lower elevation winter ranges, and are consequently exposed to human harvest. Thus, changes in the pattern of human harvest could potentially explain the decline in GYE elk since 1995. While human harvest does contribute to elk mortality, harvest levels in the GYE have declined substantially since 1995, rather than increasing. General hunting season quotas established by the Montana Department of Fish Wildlife and Parks have been reduced for Elk Management Units in the GYE, and late season population control hunts of antlerless elk in units directly north and west of the park have been reduced by more than half, or closed altogether. This pattern of reduced harvest is in contrast to many Elk Management Units outside the core wolf recovery area, where quotas have been liberalized, hunting seasons have been extended, and the state is actively promoting increased harvest to limit ongoing elk population growth (at a statewide rate of 2.8% annually).

▲ PHOTO 5.3 Elk consume large amounts of energy as they use their hooves and muzzles to dig craters in loose snow to expose dry grass and leaves. When the snow gets too deep or develops a layer of hard crust, they are likely to shift their feeding to less nutritious woody twigs. When wolves are present they relocate into timber to avoid detection, again shifting to a diet with a high proportion of woody browse. (Scott Creel)

Grizzly bears are capable of killing adult elk, but they obtain meat mainly by scavenging winter-killed elk when they emerge from hibernation and by predation on newborn elk. In the first few weeks of life, elk calves remain stationary and hidden while the mother is away, and bears are the most common predator of Yellowstone elk during this 'hiding' period. Grizzly bears have been increasing in the GYE over the period of wolf recovery. For the ecosystem as a whole, it is likely that the limiting effect of bears is stronger now than it was when wolves were reintroduced in1995. However, this change in bear numbers began many years before the elk trajectory shifted from growth to decline, and changes in bear numbers since 1995 have been relatively small in comparison to the 12-fold increase in the number of wolves. Moreover, the increase in the size of the GYE bear population is mostly due to geographic expansion and increased bear numbers on the periphery of the ecosystem, while grizzly bear density in the core of the ecosystem (where elk have declined most) has changed little, if at all, since 1995. In addition, Yellowstone wolves have largely been specialists on elk, while grizzly bears are omnivores whose diets include many elements other than meat. For these reasons, grizzly bear predation is not likely to have increased enough to be a large driver of the changes in elk population dynamics.

To summarize, GYE elk populations have declined substantially since 1995, while wolves have increased by a factor of 12. The observed decrease in elk numbers was larger than expected, and is not well-explained by ecological limiting factors other than wolves. These observations raise an interesting question that must be answered if ecology is to become a better predictive science: why was the effect of wolf predation on elk dynamics larger than anticipated?

Direct Predation and Risk Effects

Why was the observed effect of wolf recovery on elk dynamics larger than anticipated? To address this question, reconsider the Park Service's summary of the pre-reintroduction environmental impact statement:

Fifteen North American wolf experts predict that 100 wolves in

Yellowstone would reduce the elk by less than 20%, 10 years after reintroduction. Computer modeling of population dynamics on the northern winter range predicts that 75 wolves would kill 1,000 elk per winter...

In my opinion, 'kill' is the single most important word in this statement, because it reveals the logical structure of the mathematical models of predation that were used to evaluate the likely impact of wolves on elk. In essence, these models assumed that the population growth rate of elk would depend on the population's size (with competition for food slowing growth as the population increased), minus some number of individuals that were eaten by wolves. At first glance, this seems a very reasonable way to incorporate the effects of predation on the dynamics of prey. However, this logic is incomplete in a subtle but important way. Predators do not affect their prey only by killing them. Predators also affect prey by inducing changes in their behavior. When predation risk is low or absent, prey move through the landscape and harvest food in one way. When predation risk is high, most prey species modify their behavior, and the constraints that predators place on their behavior can carry costs in terms of survival or reproduction. For a broad set of prey species, behavioral responses to predation risk include changes in habitat use, diet, movement patterns, grouping patterns, increased vigilance levels and reduced foraging time.

A large body of experimental and observational research shows that these behavioral responses are induced by an increase in the risk of predation. Some research has shown that these responses are effective in reducing the rate of predation, though this point is not as well demonstrated. Nonetheless, logic suggests that the primary benefit of anti-predator behavior is to reduce the rate of predation, and any such benefit is automatically taken into account by field studies that measure the rate of predation. For example, if elk reduce their vulnerability by moving into wooded habitats to avoid detection, then the predation rate that is measured in the field will reflect this effect, even if the researcher isn't aware of the habitat shift. In contrast, the costs of anti-predator behavior are far more subtle and difficult to demonstrate and quantify. To extend the example, a shift into wooded habitats may reduce predation, but it might also carry a cost through reduced access to preferred feeding sites. If one does not design research carefully to consider such a cost, it is easily missed or attributed to causes other than predation. Because the costs of anti-predator behavior are not obvious and are difficult to measure, they have mainly been studied in experiments with invertebrates or small vertebrates in controlled settings. These costs are usually not considered in analyses of vertebrate predator-prey interactions, and they were not considered in pre-release assessments of the likely impact of wolves on elk dynamics.

▲ PHOTO 5.4 The direct impact of predation is obvious to detect and easy to understand. When wolves prey on elk, adult males are killed more often than would be expected if elk were selected randomly with respect to sex. Risk effects, or the costs of antipredator behavior, are more subtle to detect, but also have a strong effect on elk demography and population dynamics. (John Winnie)

Recent reviews of studies with invertebrate predator-prey systems suggest that the costs of anti-predator behavior, or 'risk effects', can affect the dynamics of prey just as strongly as direct killing itself. In other words, changes in prey behavior, habitat selection, foraging patterns and diet can alter the survival or reproduction of prey just as much as direct predation itself, or even more. When risk effects occur, it is a serious oversimplification to model the impact of predation simply by subtracting out the number of prey animals that are directly killed. If risk effects are important in large vertebrate systems like wolves and elk in the GYE, then risk effects might explain

▲ PHOTO 5.7 Elk are more vigilant on days that wolves are locally present within a drainage, and consequently forage less. This response is driven entirely by the responses of females. Bulls are not more vigilant in response to wolf presence, and consequently do not reduce their feeding time. (U.S. Fish and Wildlife Service)

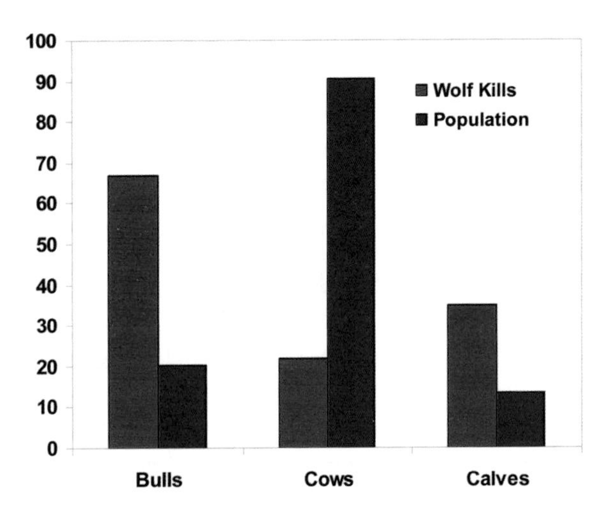

▲ FIGURE 5.1 Bull elk are killed by wolves more frequently than expected, based on their representation in the population. Cow elk are killed less often than expected. Calves, like bulls, are killed more often than expected, but only after the first six months of life (these data are restricted to winter, when calves are 6-10 months old). (Scott Creel, Dave Christianson & John Winnie)

This sex difference does not arise because bulls are less vulnerable to wolves. In fact, direct data show that bulls are killed more often than expected by chance, while cows are killed less often than expected. Instead, it appears that bulls respond weakly to predation risk because they are less able to pay the energetic costs of reducing their foraging time. The fat content of bone marrow is a sensitive measure of the degree of starvation, because fat stores in bone marrow are among the last to be depleted when an animal is running a negative energy budget. Bull elk enter winter in a depleted condition, in comparison to cows. Following their exertions in the fall rut, the marrow fat of bulls in early winter is depleted to levels typical of cows at the end of winter. This is an interesting example of differences between the sexes in behavior being driven by variation in the costs (starvation risk), rather than the benefits (reduced predation risk), which tend to be considered first.

HERD SIZE

For species that are typically found in herds or flocks, it is generally argued that shifting into larger groups should reduce the risk of predation. This can occur for two basic reasons. First, larger groups may be better able to detect or deter predators. In this situation, the risk that any member of the group will be killed decreases as the group gets larger. This is known as the 'many eyes' hypothesis. Second, it is possible that groups are no better are detecting or deterring predators, or that these benefits are offset by an increased likelihood that predators will find and attack larger groups. In this case, the risk that someone in the group will die may hold constant or even increase as group size increases, but this risk is 'diluted' among a larger number of individuals. To illustrate, natural selection should favor prey who choose to be in a group of five victims in 1000, rather than a group of two victims in 100. This is known as the 'dilution of risk' hypothesis.

▲ PHOTO 5.8 Changes in elk distribution are associated with changes in behavior and habitat selection. On days that wolves are not present, elk are likely found in open grasslands and sage meadows. When wolves are present, they move into woodland edges and are less detectable when local predation risk is high. (Ken McElroy)

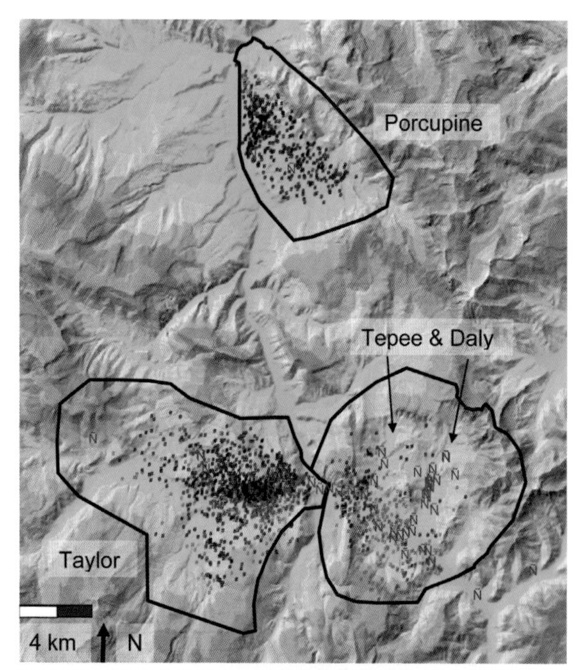

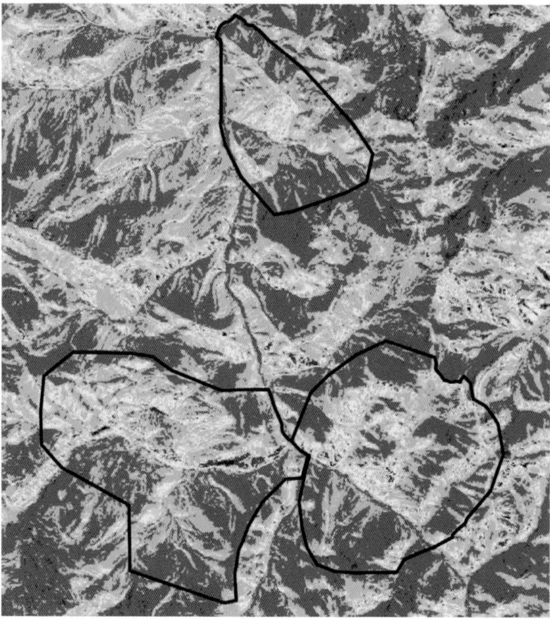

▲ FIGURES 5.2 The top image shows habitats on the Gallatin Canyon winter range, color scaled with grassland in areas of red and yellow, and coniferous woodland in areas of purple and blue. The bottom image displays the four drainages that form the three main winter ranges of elk in the Gallatin Canyon outlined in black, with the locations of wolf killed elk (blue crosses) and the locations of GPS-collared elk when wolves were present (red) and absent (black). Elk were more likely to occupy wooded areas on days that wolves were present, leaving their preferred foraging sites in open areas dominated by grasses. (Scott Creel & John Winnie)

Given the well established benefits of many eyes and dilution of risk, the general expectation is that herd size should increase in response to predation risk. It is consequently somewhat surprising that Gallatin elk formed significantly smaller herds on days that wolves were present, in comparison to the same elk on days that wolves were absent. While our research did not directly establish the function of breaking into smaller herds, a logical suggestion is that smaller herds might be less detectable. This interpretation is reinforced by data which I discuss below.

HABITAT SELECTION

In the Gallatin, wolf kills were more common in grassy areas far from woodland edges than in areas closer to cover. This pattern was also observed in the Northern Range, where kills were most common in flat, grassy areas far from timber but close to rivers. Elk prefer open grassy meadows when wolves are absent, but in response to the patterns just described, move into coniferous woodland when wolves are in the area. In our data, this response could be seen in two ways. First, we recorded the locations of all elk herds that we spotted while walking fixed transects - herds were much more likely to move far into the open on days that wolves were absent. Second, we used radio collars with onboard GPS units to record more than 20,000 locations from elk that were sampled at two hour intervals, around the clock, for two years. The GPS data also showed that elk are substantially more likely to be in wooded locations when wolves were present, and more likely to be in grassland areas when wolves were absent.

DIET AND NUTRITION

Parallel to shifts in habitat selection, the diets of elk change when wolves are present. Female elk browse on woody vegetation more and graze on grasses less. Males show less pronounced responses. The dietary shift provoked by wolves affects the quality and quantity of food that elk obtain. Surprisingly, the quality of the diet actually improves in response to wolves (higher nitrogen content, no decrease in digestible energy content). However, the quantity of food obtained goes down, and this effect is large enough to overwhelm the change in quality. The net effect of wolf presence in winter is an increase in the rate of body mass loss due to changing feeding habits.

Measuring Risk Effects on Reproduction and Population Dynamics

One of the biggest challenges for field research on risk effects is to document a causal chain from behavioral responses to risk, to physiological or energetic costs of these responses, and then to changes in survival or reproduction that affect population dynamics. For risk effects to be important, this chain must exist, but very few studies in the wild have examined every link in the chain. Consequently, an important final stage of this study was to test whether the responses we detected were associated with changes in elk demography and dynamics.

We first addressed this question by measuring progesterone levels, using non-invasively collected scat samples. Progesterone is a steroid hormone secreted by the ovaries, and in all mammals, progesterone levels increase dramatically during pregnancy, particularly

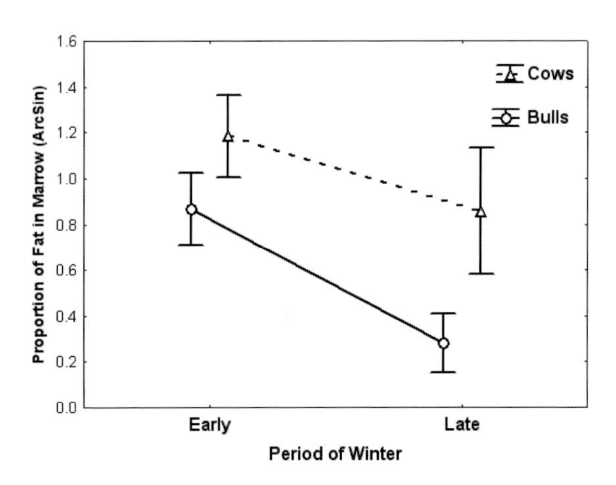

▲ FIGURE 5.3 Bull elk have lower fat stores than cow elk during winter, due to fat depletion as a consequence of competition for mates during the fall rutting season. Bulls enter winter with fat stores comparable to those of cows in late winter. Because of more limited energy stores, the antipredator responses of bulls are more highly constrained than those of cows, due to the real prospect of starvation before the spring green-up. (Scott Creel, Dave Christianson & John Winnie)

during the third trimester. Progesterone is cleared from the blood by the liver and passes into the feces intact, so measurements of progesterone in scat samples can be used to determine whether or not a female is pregnant. We collected fecal pellets from elk on five winter ranges between 2002 and 2006, and used immunoassays to measure progesterone concentrations for 1465 samples collected between March 15 and May 15, in the third trimester of gestation. When we examined the relationship between the mean progesterone level for a population and the level of predation pressure (measured as the elk-wolf ratio), we found that progesterone levels were dramatically lower in populations with high wolf-elk ratios. We then tested whether this physiological response was associated with calf production, and found that progesterone levels were a good predictor of calf numbers the following year. These results can be combined to show that calf production declines rather strongly as predation pressure increases. For these populations calf production is a good predictor of changes in population size.

Prior to describing our research results, I stated that:

GYE elk populations have declined substantially since 1995, while wolves have increased by a factor of 12. The observed decrease in elk numbers was larger than expected, and is not well-explained

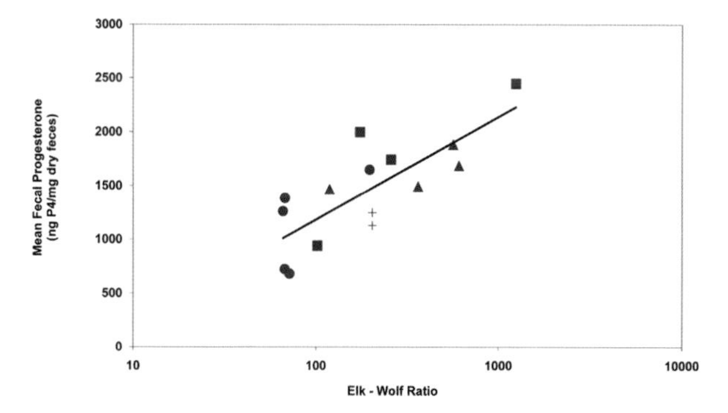

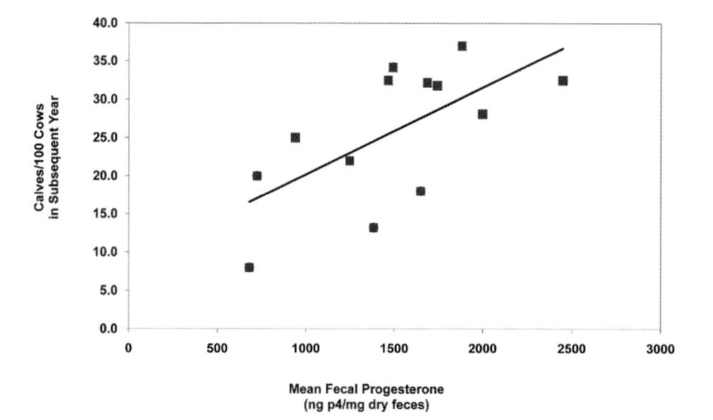

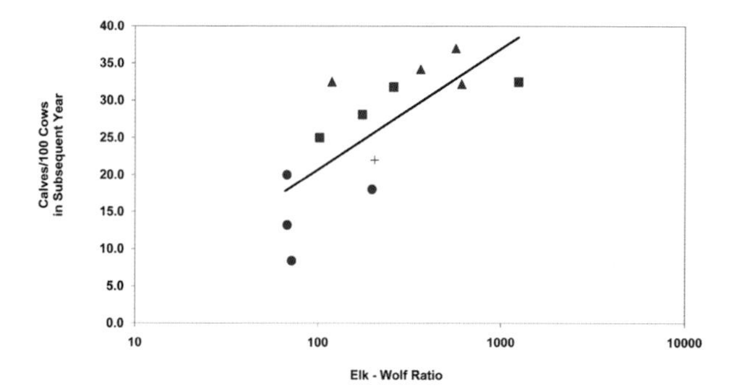

▲ FIGURE 5.4 There are clear relationships between predation pressure (as measured by the ratio of elk to wolves), progesterone levels (an indicator of pregnancy), and calf production. (top) Across populations and years, progesterone levels are higher in populations with high elk:wolf ratios. (middle) In turn, progesterone levels are related to calf recruitment. (bottom) Consequently, calf production is significantly lower in populations with greater predation pressure. Overall, populations with high per-capita levels of predation had lower progesterone levels and produced fewer calves. While the possibility is rarely considered, these data show that predation can affect prey numbers by reducing reproduction, not only by decreasing survival. (Scott Creel, John Winnie, Dave Christianson & Stewart Liley)

▲ PHOTO 5.9 This Yellowstone bull elk is in its prime. The real impacts on elk populations from wolves are complex and include direct predation as well as behavioral changes. Understanding the ecological and demographic consequences of these behavioral responses is an important part of a complete understanding of wolf-elk dynamics. (U.S. Fish and Wildlife Service)

by ecological limiting factors other than wolves. These observations raise an interesting question that must be answered if ecology is to become a better predictive science: why was the effect of wolf predation on elk dynamics larger than anticipated?

To return to this broad question, one major reason that the impact of wolves on elk dynamics was greater than anticipated is that pre-release assessments considered only direct predation, and ignored risk effects. Subsequent research has revealed that elk, like most prey species, engage in a broad set of behavioral responses to risk, that these responses carry nutritional and physiological costs, and that these costs are associated with a decrease in calf production that helps to explain the observed impact of predation on elk dynamics. Retrospectively, these results help to explain wolf-elk dynamics in the GYE. Prospectively, they suggest that we must broaden our analyses to include risk effects, if we are to accurately predict or measure the impact of predators on prey dynamics.

Chapter 6

Brucellosis in Cattle, Bison, and Elk: Management Conflicts in a Society with Diverse Values

Chapter 6

Paul Cross and his colleagues have studied not just the biology of brucellosis but the politics as well – the two seem to be inseparable. Before coming to Montana, Paul worked with bovine tuberculosis in the African buffalo of the Kruger National Park in South Africa but nothing prepared him for the agenda-driven politics of cattle ranching, bison, and big game hunting in states like Montana and Wyoming.

Not all threats to the ecology and integrity of the park are as easily visible as fire or human population growth. The issue of brucellosis in the park is a good example. Brucellosis is a bacterial pathogen found in bison, elk, and domestic cattle that can cause the host to abort its fetus. USDA successfully eradicated brucellosis in the U.S. cattle industry in 2000. Today, bison and elk are vectors that could return the disease back to domestic herds. Needless to say, the politics of cattle production and wildlife management pervade any discussion of controlling or managing the disease. If brucellosis is consistently found in cattle herds in the region, the U.S. Department of Agriculture's Animal and Plant Health Inspection Service (APHIS) amends its brucellosis regulations and require cattle producers to test their animals for brucellosis prior to interstate movement. This adds cost and logistics to their business.

Cascades of issues follow the brucellosis debate. Bison are a dominant symbol of the West yet; government officials kill them as if they were vermin. At the same time, elk, a high value public commodity, receive preferential management treatment by state fish and game agencies. Public feeding grounds are proven sources for disease but continue to be funded by the State of Wyoming, the entire debate is framed as an "old west/ new west" lifestyle choice.

Paul and his team of collaborators combine field data with mathematical modeling and statistical analysis to bring a better understanding of the pathogens to policy makers. In a cultural setting where cattle production and free roaming native species are held in equally high esteem, the science will always be political but that doesn't exclude the need for rational and effective management strategies. Paul has his work cut out for him.

J. Johnson

Mature bull bison in YNP (Steve Hinch)

Chapter 6

Brucellosis in Cattle, Bison, and Elk: Management Conflicts in a Society with Diverse Values

Paul C. Cross, Mike R. Ebinger, Victoria Patrek, and Rick Wallen

Paul C. Cross, U.S. Geological Survey, Northern Rocky Mountain Science Center,
Bozeman, MT 59715; Email: pcross@usgs.gov

Mike R. Ebinger, Big Sky Institute, Montana State University,
Bozeman, MT 59717; Email: mrebinger@hotmail.com

Victoria Patrek, Department of Ecology, Montana State University,
Bozeman, MT 59717; Email: vickipatrek@hotmail.com

Rick Wallen, National Park Service, P.O. Box 168,
Yellowstone National Park, WY 82190; Email: rwallen@nps.gov

Paul's work is highlighted on the USGS site at: http://www.nrmsc.usgs.gov/staff/pcross/research

The Greater Yellowstone Ecosystem (GYE) abounds with charismatic wildlife, picturesque landscapes and long-standing controversies. The management of brucellosis, a disease caused by a bacterial pathogen of bison, elk and cattle that can cause the host to abort, is one such example.

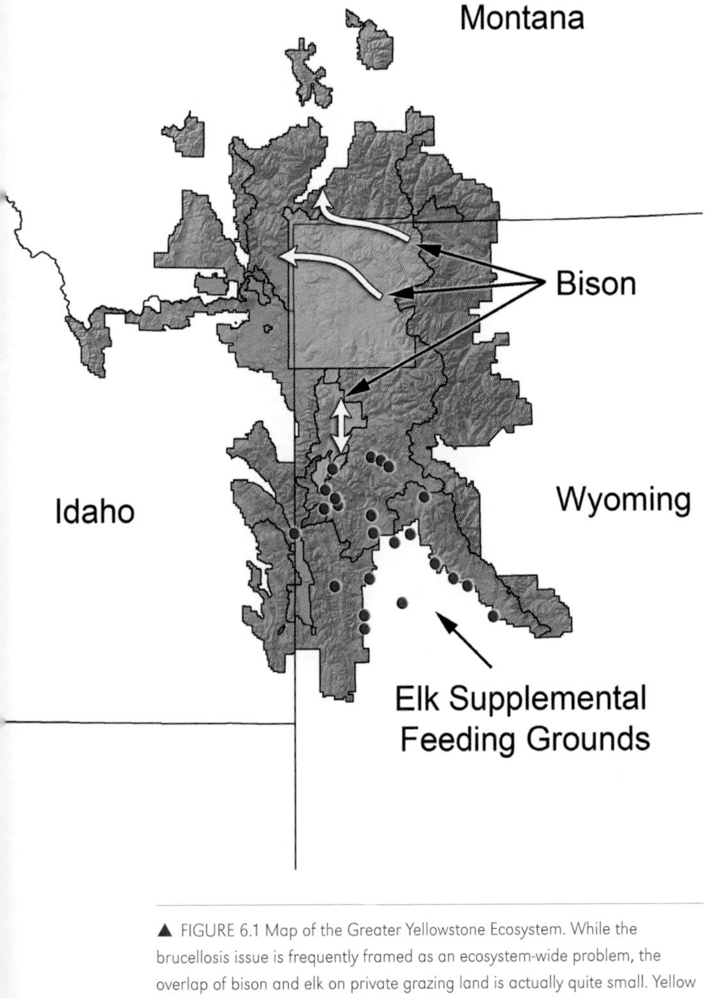

Montana
Bison
Idaho
Wyoming
Elk Supplemental Feeding Grounds

▲ FIGURE 6.1 Map of the Greater Yellowstone Ecosystem. While the brucellosis issue is frequently framed as an ecosystem-wide problem, the overlap of bison and elk on private grazing land is actually quite small. Yellow arrows show annual migrations of bison from Yellowstone and Grand Teton National Parks shown in green to public lands (brown). The 23 supplemental elk feedgrounds are shown as red circles. (Paul Cross, USGS)

The national goal of eradicating brucellosis from the livestock industry evolved in the 1930's and was formally established by law in the 1950's. Since that time the eradication program made impressive progress, and by 2000 only a small number of infected herds remained. In February 2008, the U.S. Department of Agriculture announced that, for the first time in history, the United States cattle herd was brucellosis free. The excitement was short-lived however, because cattle in both Montana and Wyoming were subsequently infected. Both states lost their brucellosis-free status, resulting in additional testing requirements and interstate movement restrictions.

The brucellosis controversy is jurisdictionally complicated, involving a variety of federal and state agencies with very different mandates. State wildlife agencies have jurisdiction over wildlife, while state livestock agencies regulate livestock movements among landowners to control the spread of infectious diseases. However, the U.S. Forest Service and Bureau of Land Management (two federal agencies that reside in different departments—the U.S. Department of Agriculture and U.S. Department of Interior) control most of the public lands that are used by wildlife and livestock in the West. On these lands the Forest Service and Bureau of Land Management must balance the interests of livestock grazing with wildlife protection and outdoor recreation. To make things even more complicated the U.S. National Park Service and U.S. Fish and Wildlife Service have complete autonomy over their lands and most wildlife management within national parks and national refuges, respectively.

Since wildlife move freely across jurisdictional boundaries, management responsibilities are shared among state and federal agencies, and conflicts arise due to their differing mandates and management philosophies. With respect to brucellosis the stakes are high, and a cutthroat atmosphere arises as the wildness and conservation of a species is pitted against the livelihood of an industry and way of life. Ultimately, the conflict revolves around the cultural and scientific suitability of management actions. The efficacy of different management strategies is usually unknown, and for some people the implementation of those strategies (e.g. capturing and hazing bison) diminishes the wild aesthetic of the species and landscape. As long as elk and bison are infected with brucellosis they represent a disease risk to cattle, but the magnitude of those risks and whether the management actions are justifiable are intensely contested.

Research in the Greater Yellowstone Ecosystem is often driven by the scientific, economic, and political context of the time. So, to understand why researchers are tackling particular issues we first develop some of the background around the brucellosis issue prior to delving into active research projects. Research is about solving mysteries, and in that spirit we raise a number of conundrums throughout the chapter. Some of these we attempt to answer, but many are open questions that researchers continue to work on.

Discovery and Detection of Brucellosis

Brucellosis has a long history with humans and their domestic counterparts. The first known record of brucellosis in humans dates back to 1859 by Jeffrey Allen Marston. His accounts were of a mysterious disease, now believed to be brucellosis, infecting soldiers of the Crimean War. In fact, during the same year Florence Nightingale returned unwell to England from the Crimean War where she had set up a hospital to treat sick soldiers. She remained chronically infected until her death, and it is believed to have been brucellosis, then called Mediterranean fever.

Captain David Bruce was sent to the island of Malta to study the mysterious fever. By chance Bruce (and others) discovered that it was the goats' milk fed to patients that was responsible for the transmission to humans. Bruce and his coworkers isolated a bacteria and ultimately it was Bruce's name that was forever attached to the causative agents. It was Danish professor L.F. Benhard Bangs who isolated a different causative organism in cattle in 1895, giving it the name *Brucella abortus*. Nomenclatures shifted, and for a brief period brucellosis was referred to as Bang's disease.

Accurate diagnosis of disease is tricky, even for human diseases. For example, recall your last tuberculosis test. The doctor probably injected a small amount of fluid into your arm. You then returned to the doctor's office 48 hours later. If there was swelling, then your immune system reacted to the injection indicating that you were previously exposed to tuberculosis. The test does not indicate the extent of the infection, when it may have occurred, or whether you have already recovered. Many tests for other diseases are similar in that they are often based upon the presence or absence of antibodies. If antibodies are present then your immune system has seen the particular pathogen in the past.

In the case of brucellosis, researchers sacrifice elk and bison that have *B. abortus* antibodies in order to determine the relationship between positive test and the extent of the infection. Tissue samples from the slaughtered individuals are taken into the laboratory, and placed into petri dish environments that promote bacterial growth. If *B. abortus* appears in the petri dish, then the animals are referred to as "culture positive." Roughly one half of the elk and bison that test positive for *B. abortus* antibodies are actually culture negative. Individuals may be culture negative because they have recovered from a previous infection or because the researcher did not capture the *B. abortus* bacteria in the tissue samples. Either way, most researchers believe that these culture negative individuals are unlikely to pass the disease to others either because they are truly recovered or because the infection was not very severe.

▲ PHOTO 6.1 Bison roam large distances in herds that vary from tens to hundreds. This herd, in the Lamar Valley, YNP, is feeding on grass and sedges in the sagebrush meadows. When bison migrate out of the park in search of winter forage, the brucellosis problem increases in complexity as bison move onto private property where state agencies have regulatory jurisdiction. (NPS, Yellowstone National Park)

Brucellosis Biology

Brucellosis in the GYE is the result of infection by the bacteria *Brucella abortus. Abortus* refers to the way that this bacteria gravitates towards the reproductive tissues of an infected host where it multiplies and sometimes causes a host to abort a pregnancy. If an infected female aborts, or even if she has a live birth, the fetus and/or associated fluids and tissues are highly contaminated with infectious bacteria. If other animals investigate those infected materials they may also become infected and pass on the infection during their next pregnancy. From a management and epidemiological perspective, brucellosis is only of concern with females as males are not considered effective transmitters of infection. Although *Brucella abortus* pathology differs among cattle, elk, and bison, generally newly infected individuals are likely to abort their calves in the first few years after the initial infection. Afterwards they recover and presumably raise successful offspring.

Brucellosis probably causes a minor decrease in the population growth rates of elk and bison, but it is not currently considered a threat to their long-term survival. In fact, many (but not all) of the bison and elk populations in the GYE are larger than they have been in the past 30-100 years. Brucellosis also does not appear to threaten the survival of individual cattle, but it can infect people and infected cattle produce fewer viable calves so the USDA judged it more beneficial to control/eradicate the disease by depopulating infected ranches. These management-related depopulations can have large impacts on the small number of affected farms (less than 10 on the Montana side of the GYE) that also serve to maintain open-space in an area of rapid human development.

Bison and Elk as Wildlife Disease Hosts

Brucellosis remains a problem in the GYE despite the overwhelming success of the brucellosis eradication program in cattle because the disease is maintained independently in both elk and bison. Brucellosis was first detected in the Yellowstone bison population around 1917 when blood collected from two female bison that aborted their pregnancies at the Buffalo Ranch tested positive for the disease.

Yellowstone bison had numerous opportunities to contact the disease from potentially infected cattle during the early historic period of the park. Prior to 1917 cattle were routinely kept in the park for milk and beef production to feed park visitors and staff. Early bison caretakers used milk from domestic cattle to feed orphaned bison calves before they were released to mingle with the rest of the herd. Yellowstone bison currently have a brucellosis seroprevalence of around 50%.

Every year in late winter as the snow piles up in Yellowstone National Park (YNP), bison migrate to low elevation winter ranges outside the Park boundary where less snow makes foraging easier. Bison that migrate out of the park encounter a landscape where cattle ranching activities conflict with bison conservation near West Yellowstone and Gardiner, Montana. Once bison have left YNP they enter the jurisdiction of Montana Fish Wildlife and Parks (MFWP) and the Montana Department of Livestock (MDOL), which have different constituencies and mandates. MFWP treats the animals as a game species, while the MDOL view them as threats to the livestock industry. To manage bison in the conflict zone, these agencies, along with YNP, the Gallatin National Forest and the U. S. Animal and Plant Health Inspection Service developed an

Interagency Bison Management Plan (IBMP) in 2000. The intention of this plan is to "maintain a wild, free-ranging population of bison and to manage the risk of brucellosis transmission from bison to livestock in Montana". The plan is focused on making sure that bison and cattle are separated during the late winter and early spring when the transmission of brucellosis is most likely. The IBMP allows for some bison in designated management areas during portions of the year that risk of brucellosis transmission is low. The plan calls for more aggressive control and culling of the population as the risk increases. Managing for a population abundance of about 3000 bison was determined to minimize the risk of bison migrating beyond the park boundary and thus reduce the risk of brucellosis transmission from bison to cattle. To keep bison within designated management areas and to keep abundance in these areas within accepted limits the agencies use a variety of tactics (riders on horseback, snowmobiles, helicopters) to haze bison away from cattle occupied areas. If necessary, they use corral traps located in the Madison Valley and Gardiner Basin to capture bison and remove them from the population.

In 2008, 1729 bison were removed from Yellowstone through hunting and management actions, roughly 40 percent of the pre-winter population estimate. This was the largest removal in the history of YNP. Conservation groups vary in their approach and philosophy, but most objected to this level of removal and the way in which it occurred. Part of the controversy revolves around the appropriate use of public lands outside of YNP. Some believe that bison, like other wildlife species, should be allowed access to public land, but this potentially brings them into close proximity with cattle herds. The extensive press coverage of bison management activities suggests that bison are a major risk of transmission to cattle. In fact, as is often mentioned by the press, there are no confirmed cases where bison have transmitted brucellosis to cattle in the wild. This is true, but not because bison are unable to transmit the disease to cattle, rather it is because the current management practices of hazing, boundary quarantines, and removal effectively separate cattle and bison. The management regime is unpalatable to many conservation groups, but it is highly effective.

▲ PHOTO 6.2 In order to manage the bison population and prevent the spread of brucellosis to livestock, public lands agencies frequently haze bison away from cattle grazing areas using aircraft, snowmobiles, and horses. If the animals cannot be kept away from cattle they are often trapped and removed from the wild population. (Buffalo Field Campaign)

Determining the source of infection when cattle test positive is a difficult problem. The events are extremely rare and detection can be anywhere from months to years after the infectious event occurs. State wildlife veterinarians use information on cattle and wildlife commingling, as well as genetic tools to determine the most likely cause of an infectious event. In all the recent cases of cattle that tested positive for brucellosis in Montana, Idaho, and Wyoming experts have pointed to elk as the most likely source of infection.

Elk are ecologically, behaviorally, and epidemiologically different from bison, and these differences present substantial challenges from a disease management perspective. Elk require an alternative set of tools than those used for bison disease management. For example, elk numbers and behavior prevent managers from using hazing as a management tool. Elk show lower disease prevalence than bison, but the prevalence in elk varies geographically. The prevalence of brucellosis in elk is higher in the southern regions of the GYE than in the north. This geographical difference in elk prevalence is due to another controversial management strategy - supplemental feeding.

In the Jackson and Pinedale regions of Wyoming, state and federal wildlife managers feed elk during the winter at 23 sites to control the spread of brucellosis from elk to cattle. The supplemental feeding program cost the

Capturing Elk and Bison

Blood samples are necessary to determine brucellosis seroprevalence, getting these samples require that elk be captured. The two capture methods used on elk and bison are the corral trap and remote delivery darting. Corral traps are used to capture a large number of animals whereas darting is used when only a few individuals are targeted. Elk are baited into corral traps with hay. After after several dozen elk are in the corral a door is released, trapping them. Bull elk are excluded from the corrals by vertical bars that are too narrow for their antlers to pass through. Bison are more easily herded using horseback riders to direct groups of animals through the opening in the corral trap. The captured animals are then coerced through a series of progressively smaller pens to a series of chutes until a single animal is contained in very tight quarters. At this site age, sex and morphology information along with a blood sample is collected.

When using darting techniques, the target individual is identified and shot with a tranquilizer dart from a CO_2 powered gun. Within a few minutes the animal succumbs to the sedation process and lies down on the ground. The capture team monitors the breathing and heart rate of the animal while data such as sex, age, weight and tissue samples are collected; a reversal drug is administered and shortly thereafter the animal is up and walking (or running in some cases) back to the safety of the herd.

Research is also being done on the feedgrounds that does not entail capture. Remote cameras are being used by Wyoming Game and Fish Department to look at how often elk come in contact with non-infected fetuses, and how quickly scavengers remove these fetuses from the feedground. This is providing managers with information on how changes in the feeding regime may decrease contacts with infective tissues and how an intact scavenger community may help reduce transmission. Researchers also use fecal samples collected off the ground to look at stress hormones. They have found that stress hormone levels in elk on feedgrounds are much higher than free-ranging elk. However, they were unable to determine what caused these high stress hormone levels. Currently researchers are investigating what factors contribute to these high stress hormone levels and are addressing how management may mitigate these high stress levels by altering feeding procedures.

state of Wyoming $1.5 million in 2007. Unfortunately, the feeding also appears to increase the prevalence of brucellosis among the portion of the elk population that frequent the feeding grounds. This leads us to another riddle. Why do managers spend time and money on a policy that increases the prevalence of a disease in one host in order to decrease the chances that it infects another?

Elk on native ranges are less effective hosts for brucellosis than bison because they often have their calves in seclusion and clean up any afterbirth as an anti-predator strategy. This makes it unlikely that other elk contact the infectious material. However, the supplemental feedgrounds create dense aggregations of elk during late pregnancy and into the spring calving season thereby allowing brucellosis to more easily persist. As a result, the prevalence of brucellosis on the feedgrounds is much higher than in other elk populations around the GYE. Outside the GYE, brucellosis is not known to persist in elk populations. Unlike elk, bison are aggregated year-round and have their calves in closer proximity to one another, thus increasing the number of potential transmission events.

With this background we can now return to an earlier question about why managers spend time and money on a policy that increases brucellosis prevalence in elk

▲ PHOTO 6.3 Darting elk on the Wyoming feedgrounds is sometimes the only way for public land managers to collect blood samples to test for pathogens that threaten wild populations. (Vicki Patrek, MSU)

in order to decrease the chances that elk infect cattle. Essentially, managers are caught in a cycle—supplemental feeding helps to separate elk from cattle, but also increases transmission and prevalence of brucellosis in elk, requiring the continued feeding of elk.

The Wyoming Governor's Brucellosis task force acknowledged that decommissioning the elk feedgrounds would likely lead to a decrease in brucellosis seroprevalence among elk, but were concerned that reduced feeding would lead, particularly during the first few years, to increased transmission from elk to cattle. Wyoming's Game and Fish Department has not yet been willing to accept these short-term risks which would likely reap long-term reductions in elk brucellosis, perhaps due, in part, to conflicting interests to support high elk populations for hunting.

Management Strategies

Not all of the constituencies involved in the brucellosis debate have the same management goals, and the most 'effective' strategies depend upon where one sits. Strategies that reduce livestock risk may not effectively protect wildlife species and vice versa. In fact, some people argue that the best management of bison and elk would be none at all. However, regardless of disease risks, concerns about private property damage would ultimately lead to some level of bison containment within a delineated conservation area. Thus, even if managers were able to eradicate brucellosis from the GYE, there would still be some form of bison management activity necessary. There is a suite of management options that focus on maintaining spatial and temporal separation of bison and cattle (e.g. conservation easements, fencing, and alternative grazing strategies), which are important for the conservation of bison but generally do not protect livestock from the risk of brucellosis transmission from elk.

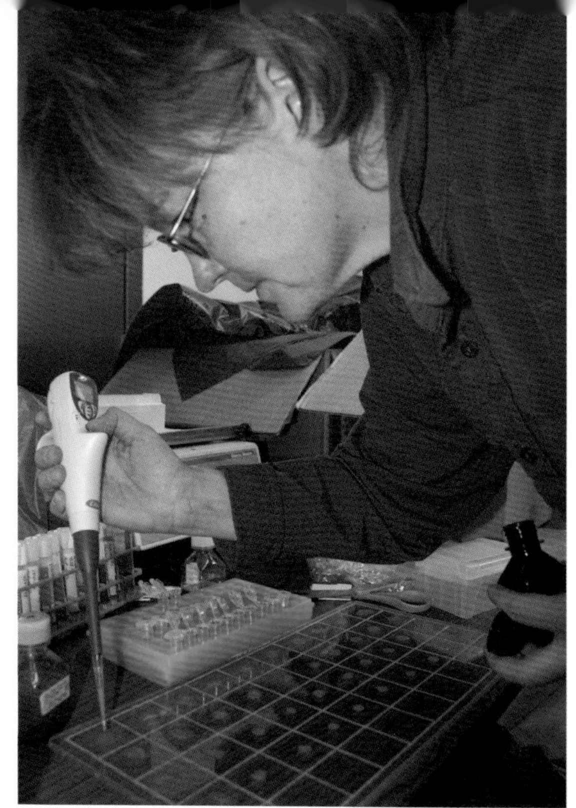

▲ PHOTO 6.4 Wyoming Game and Fish personnel test elk samples for brucellosis. Blood testing shows the proportion of animals that have been exposed to brucellosis and developed antibodies. A positive test doesn't necessarily mean that the animal can transmit the disease only that it harbors the bacteria and may be able to transmit to other animals. (Mark Gocke, Wyoming Game & Fish)

There are several reasons for the very different treatment of elk and bison in this ecosystem. First, bison congregate in large numbers more so than elk do and are thought to be more controllable. There are substantially fewer of them within the GYE compared to elk and they tend to remain in valley bottoms during much of the year. Thus, an aggressive management program to vaccinate, and/ or capture and test a high proportion of bison is believed to present a greater probability for success than the same management strategy for GYE

▲ PHOTO 6.5 Feeding time at the National Elk Refuge. Established in 1912, the refuge is the winter home of the largest elk herd in North America. Elk are supplementally fed alfalfa pellets and hay throughout the long winter. The federal and state feedgrounds were established as a partial solution to loss of habitat loss and conflicts with ranchers but have created other dilemmas. When animals are crowded together on feedgrounds they often have higher rates of disease. Closing feedgrounds could reduce the prevalence of brucellosis, but it would likely result in a different set of problems. (Paul Cross, USGS)

elk. The proposition of keeping tens of thousands of elk away from cattle is certainly more difficult than controlling a few thousand bison. Second, an established hunting constituency actively lobbies for increasing elk population sizes and hunting opportunities.

Despite the relative success of eradication efforts in cattle as well as in the bison of Wind Cave National Park and Custer State Park, eradication of brucellosis from the GYE seems unlikely. All of these successful eradication programs have involved capturing large portions of the populations repeatedly, vaccinating young females that test negative for the disease, and removing infected individuals. Such an effort may be feasible with bison, but the logistics of capturing tens of thousands of elk scattered across the rugged terrain of the GYE is hard to imagine. Reductions in the total number of bison and/or elk may reduce the total number of infectious individuals on the landscape, but are unlikely to lead to outright eradication.

Compared to removing bison or feeding elk, vaccination is an aesthetically appealing control strategy. However, the effectiveness of vaccination at eliminating pathogens, or even controlling them, is limited, particularly in wildlife species. In fact, the elimination of *Brucella abortus* from elk and bison through vaccination alone is not an expected outcome. This should not be surprising given that, even in humans, we have only successfully eradicated smallpox. Since wildlife species do not present themselves for vaccination like people do, the delivery of a vaccine becomes an important component of any control strategy.

▲ PHOTO 6.6 Elk, bison, and most other ungulates lick newborn young, whether it is one of their own offspring or not. If elk, bison and cattle mix during the birthing season, this instinct is a potential vector to transmit brucellosis to another animal. (NPS, Yellowstone National Park)

One benefit of the supplemental feedgrounds is that the aggregation of elk facilitates a vaccination program that began in 1985. Nearly all calves are vaccinated annually on all feedgrounds except Dell Creek using Strain 19 *B. abortus* vaccine encapsulated biobullets. These biobullets are hard, plastic, 0.25 caliber projectiles that penetrate the skin, dissolve in muscle tissue and deliver the vaccine dose. The Strain 19 vaccine reduced abortion events in captive elk from 93% to 71% during the first pregnancy, but did not reduce infection rates. Over the longer term, reduced abortion rates should translate into reduced transmission and thus lower prevalence. Unfortunately, this vaccination effort does not appear to have resulted in a major reduction in brucellosis seroprevalence on the feedgrounds.

Research on the development of more effective vaccines for wildlife and livestock is ongoing, but the chances of developing a *"silver bullet"* in the near future remain slim. The amount of research effort conducted on livestock and wildlife vaccines is relatively small compared to humans. Second, since several *Brucella* species were among the first pathogens to be developed into biological weapons they are highly regulated. Researchers must pass security clearances and the facilities used to house the pathogens must meet biosafety requirements. These requirements become increasingly difficult and expensive to satisfy when large animals are involved and only a few facilities in the U.S. can run these experiments.

Obviously it would be easier to vaccinate cattle than either bison or elk. However, there is an interesting disconnect between how vaccines typically work and the current livestock regulations. Most vaccines help individuals mount a strong immune defense after they have been exposed to a pathogen. They do not reduce the probability that an individual is exposed to an infection. So, vaccinated cattle are equally likely to be exposed to *Brucella abortus* (and test positive), but vaccinated cattle may be less likely to abort their calves following exposure to the disease. However, protecting the cattle from future morbidity does not help the rancher with respect to the state agencies who are still required to quarantine or slaughter the exposed (but less infectious) herds.

History of Bison in Yellowstone and Grand Teton National Parks

After the bison slaughter of the late 1800's, an active bison restoration program was initiated in YNP in 1902 when 21 animals were released into the Park from two semi-domestic herds. Brucellosis was first detected in YNP bison in 1917, and it is likely to have been introduced from cattle to bison just prior to that time around the Buffalo Ranch area of the Lamar Valley where a few milk cows were kept until 1919. The bison population grew from the initial couple dozen individuals in 1902 to over 1000 bison by 1930. From 1930 to 1950 Yellowstone bison were translocated to many sites throughout North America. In 1936, park managers relocated 71 bison from the northern range of YNP to the central region to redistribute bison to formerly occupied range within the park. The central herd then increased until the late 1950's. Between 1950 and the mid 1960's an active brucellosis eradication program was implemented in concert with a program to actively reduce abundance.

By the mid 1960's biologists felt that the brucellosis eradication program would nearly eradicate the bison population and made recommendations for YNP to switch to an ecological process management policy. By the mid 1990s the bison population was over 4000 individuals and increasingly likely to be moving outside the boundary of the National Park during the winter resulting in the reinstitution of bison removals, but this time the actions occurred along the borders. Several large-scale removals have occurred affecting over 1000 individuals in 1996, 2006 and 2008. Bison were also reintroduced to the Grand Teton National Park (GTNP) in 1948. This small herd was found to be infected with brucellosis in 1963, but through testing and removing positive individuals and vaccinating calves the herd was determined to be brucellosis-free in 1967. After an initial escape of these bison from a captive area nine of these bison were allowed to roam-free in 1970. In 1990 the herd numbered 123. By 2008 the herd was over 800, and managers were attempting to increase the hunting pressure on bison to reduce population abundance. Most of these bison spend the winter months on the National Elk Refuge just outside of Jackson, Wyoming where they are supplementally fed.

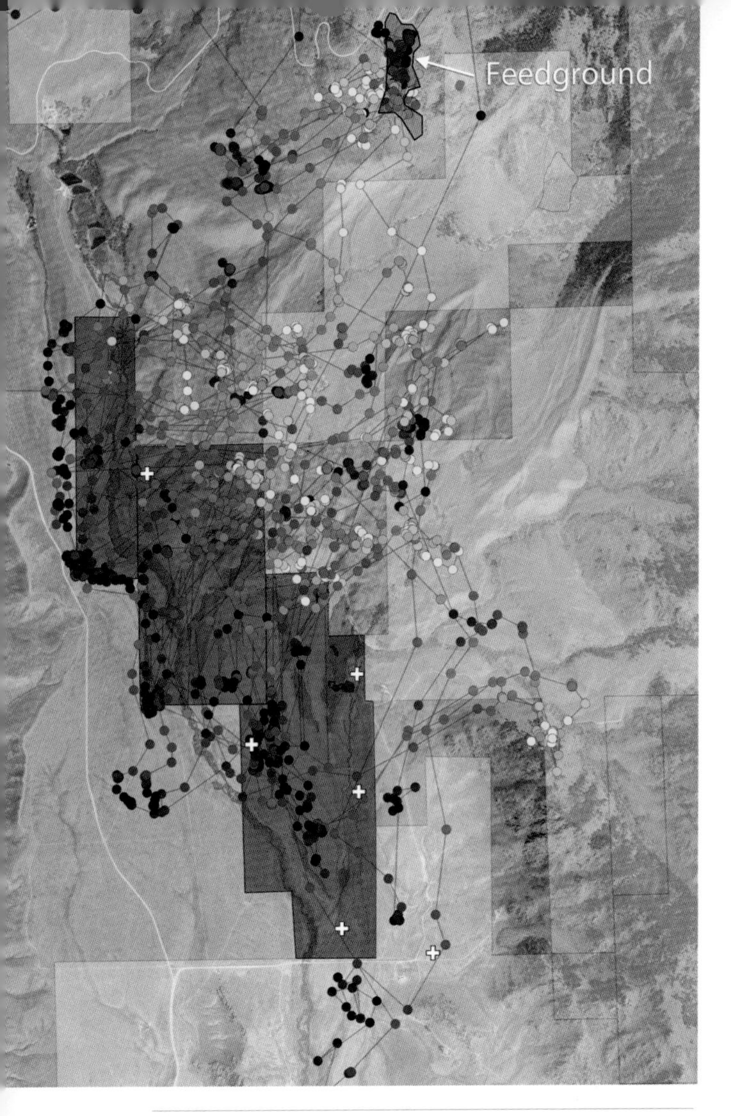

▲ FIGURE 6.2 Elk-cattle interactions may occur mostly at night. Points represent GPS locations from one elk every 30 minutes. Dark circles are during the night. Yellow circles are during the day. Red areas and white crosses are historic cattle properties and haystacks, respectively. (Paul Cross, USGS)

Current Research

Despite the importance of understanding the contact patterns of elk and cattle, very little work has been done on the issue until recently. In collaboration with Wyoming Game and Fish Department, we have been deploying collars on elk around the Wyoming feedgrounds. These collars use Global Position Systems (GPS) to communicate with satellites and record locations at specific time intervals. In our case these collars record a position every 30 minutes; after a year the collars are programmed to automatically release from the neck of the individual allowing us to collect the collar without recapturing the individual. GPS collars allow us to look at fine scale patterns in both space and time. For example, preliminary data show that elk may be going onto private land at night and concentrating on irrigated pastures indicating that elk-cattle contact may be more frequent than we might have guessed based upon daytime observations. We are also using satellite imagery to investigate how snowpack affects the artificial feeding season for elk. In previous studies we found that feedgrounds that feed longer into the spring have a higher prevalence of brucellosis. Meanwhile, the population size at the feedground appears to explain very little about the seroprevalence on the site. For example, the elk on the National Elk Refuge have the lowest seroprevalence of any feedground (around 10%) but have around ten times more elk than most of the Wyoming feedgrounds. The NER probably has a low prevalence of brucellosis because it stops feeding before most transmission occurs. Using this information Wyoming Game and Fish Department is trying to end the feeding season earlier than normal on several feedgrounds. However, there is a concern that shorter feeding seasons may result in elk moving from the feedgrounds to private properties where they may infect cattle. By combining satellite imagery data with GPS collar data we hope to generate a picture of which areas melt first and where the elk are likely to go once the feeding season ends.

The substantial support for a vaccination program within the GYE has directed much of the recent research regarding brucellosis in bison. To date, researchers have focused on three related aspects of vaccination: whether the use of strain RB51 *Brucella abortus* vaccine is safe for use in bison, whether it is safe if non-target species were to encounter the vaccine in the wild, and in quantifying any differences between vaccinates and non-vaccinates regarding their ability to mount an effective immune response to brucellosis. Research has shown that this vaccine is safe when used in bison and will not present any unusual clinical symptoms for non-target species that may encounter vaccine indirectly through exposure to vaccinated bison. However, there are differing results and professional opinions among brucellosis experts regarding the level of protective immunity that a vaccinated Yellowstone bison would exhibit. With clinical experiments completed, the next step is to conduct some experimental trials to measure the response to vaccination by Yellowstone bison in the field.

Unanswered Questions and Future Directions

"There are known knowns. These are things we know that we know. There are known unknowns. That is to say, there are things that we know we don't know. But there are also unknown unknowns. There are things we don't know we don't know." (Donald Rumsfeld Feb. 12, 2002, Department of Defense news briefing)

Although Mr. Rumsfeld was cryptically describing issues of national defense, his statement is applicable to many ecological and wildlife management issues. It is the unknown connections in ecological systems that often result in unintended consequences. For example, the release of an insect in Montana to control spotted knapweed (a non-native weed) resulted in elevated levels of hantavirus in mice. These 'unknown unknowns' are pervasive in ecology, but also troubling are the 'known unknowns'.

There are many uncertainties about how brucellosis is maintained in the wildlife of the GYE, how best to manage the risk of interspecies transmission, and whether elimination of the disease from the wildlife reservoir is technically feasible. These issues are all topics of ongoing research. In particular, transmission is very difficult to estimate in either human or wildlife systems. For brucellosis it remains unclear how often bison transmit brucellosis to neighboring elk and vice versa. If transmission between the species is rare, then the dynamics of brucellosis infection in each species are likely to be independent of one another. In other words, a decrease in the prevalence of brucellosis in bison may not result in a corresponding decrease in elk. Research on the genetic composition of *Brucella abortus* strains in elk and bison may help to unravel this question.

The management of brucellosis, like so many other issues in the Greater Yellowstone Ecosystem, is complicated by political, ecological, and economic factors. In addition to biological uncertainty, the social tolerance for ongoing intensive management through mass wasting of wildlife resources (whether elk or bison) is also uncertain. Testing to identify seropositive individuals so that they could be eliminated from populations is a proven strategy in domestic stock. Applying this type of management in a wildlife conservation arena is unlikely to occur for logistical, financial, and sociological reasons. Consequently, the social and political debate will have to resolve the issue of whether brucellosis elimination is worth the price or whether an effective risk management strategy could be acceptable with changes in the disease regulations. Such changes could benefit both the agricultural and conservation communities. The GYE is one of the fastest growing regions on the US and one where most constituents have a common goal of maintaining open space and healthy wildlife populations. Researchers and decision makers will need to continue to ask focused questions that systematically resolve scientific uncertainties. To fail to do so places the ranching and conservation constituencies arguing their own ideology unchecked by a common science based reality.

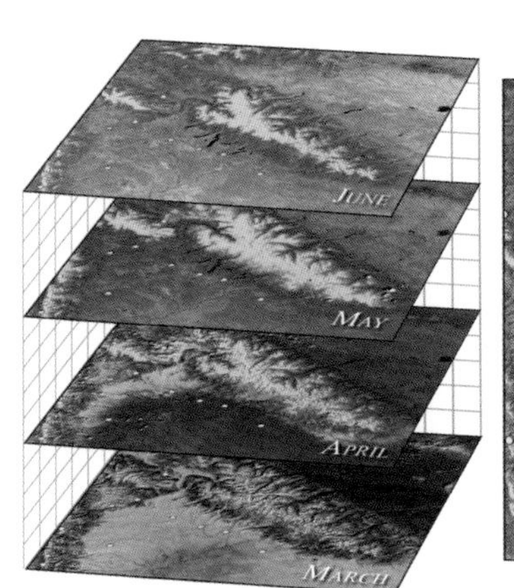

▲ FIGURE 6.3 Images of vegetation and snowcover from the Landsat 5 satellite sensors. Blue indicates snow. Red indicates soil moisture, and green indicates chlorophyll absorption (i.e. vegetative growth). Elk feedgrounds are represented by yellow circles. Images like these allow researchers to investigate how snowcover and vegetation affect the timing and routes of elk migration to and from summer ranges. (Paul Cross, USGS)

Chapter 7

Fisheries Science and Management in the Greater Yellowstone Ecosystem: Ensuring Good Fishing by Preserving Healthy Ecosystems

Chapter 7

Al Zale gets to go fishing, a lot. Al is the leader of the Montana Cooperative Fishery Research Unit. He is the coordinator for most of the applied regional fisheries research for the State of Montana and the U.S. Department of the Interior.

Montana and the GYE has as diverse a fishery as you can find in the U.S. In the west, the large pristine rivers that drain the Rocky Mountains offer unparalleled cold water trout fishing. Fly fishermen from around the world seek out the native Cutthroat, Rainbows, and Browns of the Madison, Beaverhead, Big Horn and countless other rivers managed by Montana Fish Wildlife and Parks. Fly fishing in Montana is a high value activity and part of Al's job is to see that our native fishery is sustainable. Al is particularly active in the management of Arctic Grayling in Yellowstone National Park.

In less populated (and dryer) parts of the region warm water fisheries prevail. Species like Walleye, Pike, Muskie, and Large and Small Mouth Bass, share water with Channel Catfish and Pallid Sturgeon. The prehistoric Paddle Fish is still fished in Yellowstone and Missouri Rivers. Al has responsibility for these fisheries too.

He and his collaborators have perfected techniques for capturing and tagging every type of fish in the region. He uses a range of technologies but the bottom line is to do it without harm to the individual fish or the resource. They need to assess abundance, productivity, and demographics for a research subject they can't see and is perfectly adapted to their river environment.

Fishing is the favorite pastime of visitors to Yellowstone. If a quality experience is going to be available for future generations, fish populations need to be sustainable and healthy. Al's work is aimed at ensuring policy makers have the data to make decisions that will maintain and enhance fish populations in the Yellowstone region.

J. Johnson

Size Matters (Drew Rush)

Chapter 7

Fisheries Science and Management in the Greater Yellowstone Ecosystem: Ensuring Good Fishing by Preserving Healthy Ecosystems

Alexander V. Zale

Alexander V. Zale, USGS, Montana Cooperative Fishery Research Unit, and Department of Ecology, Montana State University, Bozeman, MT 59717; Email: zale@montana.edu

Al's Montana Cooperative Fisheries Unit site is: http://www.montana.edu/mtcfru

Some of the world's best and most famous sportfisheries are located in the Greater Yellowstone region. Many of the region's clear, cold streams, rivers, lakes, and reservoirs harbor abundant populations of popular sportfish such as native Yellowstone (Oncorhynchus clarkii bouvieri) and Westslope cutthroat trout (O. c. lewisi) and Arctic grayling (Thymallus arcticus) as well as introduced Rainbow (O. mykiss), Brown (Salmo trutta), and Brook trout (Salvelinus fontinalis).

Well before the film, *A River Runs Through It,* heightened the popularity of fly fishing here, the quality and resulting economic and recreational values of the region's fishery resources were already well known and appreciated. For many, a life of fishing is not complete unless they have fished the waters of the Yellowstone region.

A primary societal value of angling is that it affords simple relaxation in the outdoors to lessen the stresses of modern life. It is also an opportunity for social activity and family fun. More pragmatically, angling is big business worth millions of dollars annually in the GYE. Fly shops, fishing guides, restaurants, motels, drift boat dealers, souvenir shops, and other businesses all profit directly from angler expenditures. Moreover, state fish and wildlife agencies and Yellowstone National Park are funded through license sales.

Fisheries are not only important for the recreational and economic values of fishing. Abundant and productive populations of both sport and non-game fish contribute to the ecological health of the GYE and the environmental services it provides. For example, 42 animal species ranging from water shrews to bald eagles to grizzly bears depend on Yellowstone cutthroat trout for food. The clean waters of Yellowstone are sources of drinking water for residents; the many rivers and lakes add to the quality of life in the region. Fisheries have wide ranging ripple effects on society and the environment in the GYE.

Fisheries management in the region has two primary goals. Maintenance of excellent fishing is a top priority; anglers want "big fish and lots of 'em". High quality angling requires active fisheries management to offset the intense fishing pressure and habitat alteration affecting many of our local waters. Restrictive regulations (e.g., catch and release, length and bag limits, closed seasons), habitat protection and restoration, and judicious stocking of hatchery fish (especially in reservoirs and mountain lakes) are some of the more important tools used by modern fishery agencies.

The other priority is preservation of native species to maintain ecosystem integrity and function. Considerable effort and funding is expended on Arctic grayling and cutthroat trout restoration and enhancement, including habitat restoration, extermination of non-native competitors, species reintroductions, and predator controls. Recently, lake trout, (*Salvelinus namaycush*), in Yellowstone Lake and brook trout in local streams have moved to the forefront of management dilemmas. Of course, sometimes these two priorities are at odds, such as when cutthroat trout conservation conflicts with rainbow trout fishing, but for the most part, what's good for ecosystem health is also good for fishing.

The inherent quality of the GYE's fisheries is a product of the area's climate, topography, and geology. Maintenance of that excellence is the result of careful and comprehensive management and monitoring. Dozens of state and federal biologists survey and monitor fisheries throughout the region. They enthusiastically implement protective regulations and management actions to preserve and enhance the resource. The sampling and analysis techniques described in this chapter are used in our regional survey and monitoring programs. The goal of any survey program is to monitor which species exist in a water body, how many there are, how large they are, how fast they are growing, and what their age structure is. This information can be used to assess the condition of a population and thereby provide insights into effects of human actions such as land use and climate change or, the success of fishery management strategies. This chapter explains how the information acquired by fish sampling is used and describes the sampling techniques used to gather it.

Goals of Fish Sampling

PRESENCE AND ABUNDANCE

The obvious difficulty of fishery research is that the subject is hidden in deep pools and rushing currents. Fish cannot be counted using aerial surveys; we can't set up a spotting scope and count them from afar. Documenting species presence is critical if we are to preserve native species that have become reduced in number and range in recent years such as fluvial (river dwelling) Arctic grayling, Yellowstone cutthroat trout, and Westslope cutthroat trout (which, despite their name are native to both sides of the Continental Divide). This is why we survey headwater streams throughout the GYE for presence of genetically pure populations. On the other hand, detection of non-native species can also be a priority, especially where such species would compete or hybridize with rare or imperiled natives. Examples include invasion of cutthroat streams by rainbow, brown, or brook trout and the introduction of lake trout into Yellowstone Lake. In both cases the nonnative often outcompetes or is a predator of the native.

Estimation of fish abundance is an important fishery management activity in the region. After all, all those anglers visiting the world-renowned GYE fishery want to catch as many as possible, so maintaining abundant populations is critical. Abundance information can be used to help set harvest limits, indicate the success of management actions, and identify possible problems. For example, the precipitous decline of the rainbow trout population in the Madison River in the early 1990s led to the discovery of whirling disease. Repeated estimation of abundances in specific river sections over many years using standardized techniques allowed Dick Vincent, a now retired Montana Fish, Wildlife and Parks biologist, to recognize a connection between hatchery raised fish and declining number of natives. Abundance estimation is also critical for managing imperiled species, which occur at low numbers by definition. Knowing just how low those abundances are can help decision makers set priorities for rescue efforts.

▲ PHOTO 7.1 Catch and release is the primary fishing philosophy on most waters in the Greater Yellowstone Ecosystem. It has proven to be important to maintaining a high quality fishing experience as fiy-fishing has become increasingly popular. This fisherman is trying his luck on Slough Creek in Yellowstone National Park - the place where catch and release was first implemented in 1973. (Jerry Johnson)

Catch and Release

Catch-and-release fishing is so common and accepted in the GYE today that it is easy to forget that it is a relatively new solution to an old problem. Overfishing was bemoaned in the angling literature for hundreds of years, but the rapid post-World War II increase in angling and other outdoor leisure activities caused the quality of many wild fisheries to deteriorate. There is also a long tradition of eating the fish we catch. Despite its inefficiency and cost, stocking hatchery fish was the typical solution used to augment some simple size, bag, and season regulations. Yellowstone National Park had a long history of recreational fish harvest despite a legal mandate to retain its natural resources in their natural condition. Even so, evidence of depletion was evident in wild trout fisheries in and around the Park.

A desire to eliminate expensive stocking led a Michigan fishery scientist, Albert S. Hazzard, to institute the "Fishing-For-Fun" program in some high-quality wild trout streams in 1952. All trout caught were released and angling was restricted to flies and lures to decrease post-release mortality. The goal was fun, not filets. The basic concept was that fish could be caught multiple times, thereby providing recreation for many anglers over and over again. The time was ripe for its introduction because of the developing sportsmanship, environmental awareness, and conservation ethic of some anglers. The program was soon copied by other states and known as the "Hazzard Plan" and later "Catch-and-Release". It proved to be a useful management tool for many different types of fisheries subjected to high fishing pressure and became popular throughout the U.S. In fact, voluntary catch-and-release is now common practice by many anglers, even where not required.

Catch-and-release was first implemented in the GYE in 1973 in the Yellowstone River, Slough Creek, and the Lamar River in Yellowstone National Park. Full implementation was preceded by gradual institution of progressively more restrictive regulations including gear restrictions, area closures (e.g., Fishing Bridge), and size and bag limits. The program soon expanded to other waters both in the Park and nearby, and has been a resounding success. The ethic maintains ecosystem integrity and function and provides excellent fishing to great numbers of anglers. Today, cutthroat trout in a reach of the Yellowstone River below Yellowstone Lake are caught and released an average of 9.7 times during the fishing season. A key to the success of catch-and-release however, is the availability of nearby fisheries where harvest is allowed. Montana continues to stock lakes and reservoirs to support harvest fisheries where catch-and-release is unnecessary or unpopular.

The primary approaches to estimating fish abundances are mark-recapture and depletion sampling. Mark-recapture involves two sampling occasions. Fish caught during the first sample are all marked or tagged and released unharmed; a second sample is collected to recapture some of them. As long as the same methods are employed both times, mark-recapture techniques are highly reliable.

Depletion sampling involves removing fish from a sample area and temporarily not returning them. If enough fish are removed that the catch in the next sample is reduced (and if catch is proportional to abundance, which it should be except in unusual circumstances), then the reduction in catch in the second sample can be used to estimate the original abundance. For example, if 50 fish are caught in a blocked-off stream section during a first sample and only 30 are caught in the next (using the same amount of effort), then the removal of the first 50 fish reduced the catch rate—and presumably the original number of fish—by 40% ($50 - 30 = 20$; $20 \div 50 = 0.4$). Therefore, $50 = 0.4 \times N$ or $N = 50 \div 0.4 = 125$

▲ PHOTO 7.2 The original Fishing Bridge was built in 1902 and refurbished in 1937. The portion of Yellowstone River beneath the bridge was a major spawning area for Yellowstone cutthroat trout and so was the most popular place for many park visitors to catch their first fish. Because of the decline of the cutthroat population and overfishing, the bridge was closed to fishing in 1973. It is still a popular place for observe fish in the clear waters of the Yellowstone. (NPS, Yellowstone National Park)

fish. Depletion estimates of fish in small streams are commonly made by electrofishing. Usually, at least three successive samples are taken to get multiple estimates of the reduction rate of the catch. These techniques are used when we want to track populations in a discrete fishery.

In many situations, an actual abundance estimate may not be needed. Rather, a relative measure of fish abundance may suffice, especially when comparing different places or times. Such measures are expressed in terms of "catch per unit effort" such as the number of fish caught in a gill net set for an 8-hour period or the number caught by electrofishing a 100-meter stream segment. Relative abundance can be measured more quickly and simply than actual abundance because only a single sample is taken. This technique might be used when we want to know if two streams that appear to be similar actually produce similar amounts of fish. If not, we might ask what we could do to increase fish population in the less productive habitat.

Restoring the Native Trout Fishery in Montana

In 1974, Montana Fish Wildlife and Parks did something radical at the time - it stopped artificial stocking of hatchery fish in streams and rivers that supported wild trout populations; the first state to do so. The standard policy was to supplement populations of wild trout with those from state run hatchery operations; sportsmen, businesses and fishery agencies relied on high fish numbers to attract out of state fishermen each season.

Beginning in 1968, FWP fisheries biologist, Dick Vincent, began studying the trout population of the Madison River and how it changed according to river flow on two sections of river. He noticed that on one section population didn't vary much, on the other it did. The difference turned out to be that one section was stocked with hatchery fish, the section where population changes tracked river flows (as expected) was a native fishery.

Vincent experimented with stocking previously unstocked areas and ending it in river reaches that had been stocked for years. He showed that stocking hatchery fish suppressed wild trout numbers. What seemed to be happening was that behaviors learned by hatchery-raised fish allows them to outcompete natives for food and habitat. Hatchery fish were more tolerant of the crowded, semi-sterile conditions in the hatcheries and so displaced less tolerant natives from good feeding lanes. The result was that the number of native fish declined as well as the average size because fewer grew to maturity. It took time but he finally convinced fishery managers to change decades of tradition and practice.

Vincent's electroshocking methods and willingness to challenge the accepted scientific wisdom of the day has meant that once stocking was discontinued, wild trout numbers doubled and even tripled on some rivers in the state. Today, Montana's native trout are seen as a unique fishing experience where the quality of the fish caught has replaced the quantity of fish. In Montana, thanks to good science and risk-taking fishery managers, an angler has a good chance of catching a big, wild trout and creating a memory of a lifetime.

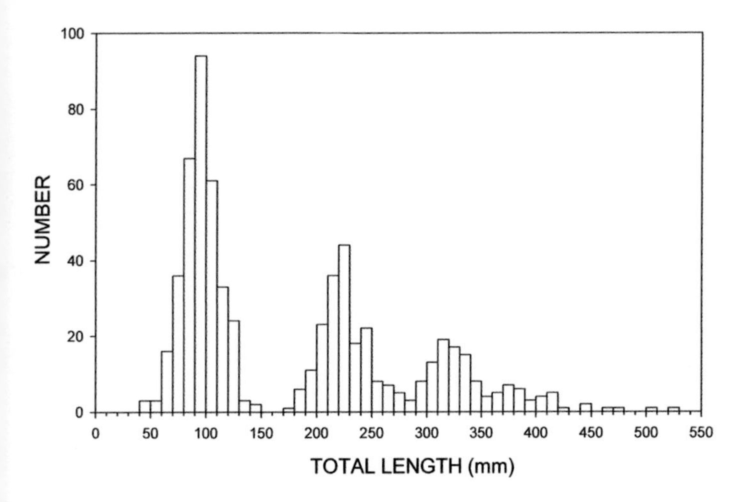

▲ FIGURE 7.1 A length frequency distribution of rainbow trout collected during spring indicates a strong cohort of age-1 fish and consistent prior recruitment. Measuring the size of the population by cohort can inform fish managers about mortality or the health of the river system. (Al Zale, MSU)

DEMOGRAPHICS AND CHARACTERISTICS

Size matters, at least to anglers. Captured fish are typically measured, normally from the tip of their snouts to the ends of their tails (i.e. "total length"). They are usually under light anesthesia induced by chemicals dissolved in the water they are being held in. Weight is recorded less frequently because weighing fish is time consuming and time out of water is stressful. Moreover, weight can usually be estimated with reasonable accuracy using existing species-specific weight-length equations. An exception is when a measure of body condition is desired. Measures such as relative weight relate the weight of a fish to its length (i.e., "plumpness") and are sensitive to environmental conditions. Plump fish tend to occur where food is abundant, water quality is good, and habitat is of high quality. Thin, "snaky" fish are indicative of poor environmental conditions, a lack of adequate food, or disease.

Length-frequency distributions depict the number of fish of each size in a sample. Natural and fishing mortality cause fewer larger fish to be present. Peaks in the distribution represent cohorts of lengths of age classes, each of which is clumped because fish hatch during a short time period. In other words, there may be many fish of the same size and age in any given sample. Gaps in a distribution may indicate years when few young fish were added to the population for a variety of reasons. Presence of few large fish may be caused by excessive harvest or food limitations. Rough approximations of annual growth rates can be made by comparing lengths between the peaks, but these tend to run together at advanced ages because of variability in individual growth. More accurate estimates require aging of fish using rings in hard body parts such as scales or otoliths - the ear bones. Integration of lengths and ages allows determination of annual growth rates. This is a useful measure of the quality of a fish's environment and also correlates well with body condition. Faster growth produces more large fish in a shorter period of time. Aging of fish also allows calculation of age-specific mortality rates. We can identify life stages when survival is poor; poor survival may be caused by poor environmental conditions, predation, or harvest and may hurt some ages more than others.

Accurately measuring the sizes of fish in a population would appear to be a relatively simple thing, but unfortunately all sampling methods and hardware tend to be size-selective in one way or another; they tend to be better at capturing small fish, or large fish, or individuals of a narrow size range. Moreover, species selectivity occurs as well. Such tendencies are described below

▲ PHOTO 7.3 Biologists can backpack an electrofishing unit into remote locations. In this case, we sampled an irrigation diversion canal to determine how many are distracted from the main river. A net was used to prevent fish from escaping from the sample reach. (Leslie Bahn)

Mark-Recapture Sampling

Ecologists and wildlife biologists frequently try to estimate populations based on sampling part of the total population. The method was developed to sample populations of animals that are highly mobile or those for which it would be difficult to know how effectively you had sampled with only a single sampling time. Mark-recapture logic is based on the premise that the ratio of previously marked fish or other animal recaptured (R) during the second sample to the total number marked (M) is equal to the ratio of the total number of fish caught during the second sample (C) to the total abundance (N):

$$\frac{R}{M} = \frac{C}{N}$$

Rearranging that equation gives:

$$N = \frac{M \times C}{R}$$

Thus, if a biologist recaptured 6 of 42 previously marked fish in a second sample of 51 fish, then N, the total abundance is

$$N = \frac{41 \times 52}{6} = 357$$

Think of it like this. Try grabbing a handful of dried beans from a jar and marking those "captured" during that first sample with a marker or replacing them with a different color bean. Return them to the jar and mix the beans thoroughly to ensure that every bean has an equal probability of being captured in the second sample. You also don't want to wait too long between samples so as to avoid any "mortality" among the beans, if say someone were to make soup. Biologists usually wait a few days between marking and recapturing to allow the marked fish to mix back into the population, but not so long that some fish could die or emigrate out of the sampled area. This procedure was first used by C.J.G. Petersen in studies of marine fishes and F.C. Lincoln in studies of waterfowl populations, and so is often referred to as the Lincoln Index or the Petersen Index.

▲ PHOTO 7.4 We can also electrofish from an inflatable raft and cover a much larger reach with less effort. This sampling is taking place on the Gibbon River in Yellowstone National Park (Amber Steed)

for each capture technique. Use of multiple gear types with different selectivity characteristics can moderate this bias. Of course, selectivity can be advantageous in situations where capture of only a limited size range of certain species is desired, as when sampling only juvenile trout to determine year-class strength.

Fish Capture Methods

Electrofishing is perhaps the most common, useful, and effective fish collection technique used in the GYE. Limited primarily by depth, it is a shallow-water collection technique useful in streams, shallow reaches of rivers, and surface waters of lakes and reservoirs. Basically, it involves creating an electrical current in the water that stuns and incapacitates the resident fish, thereby allowing biologists to scoop them up with a dip net. Electrofishing configurations range from small backpack units to shore-based units to floating electrofishers to large, powerful units mounted on inflatable rafts, drift boats, or jet boats. Power is supplied by rechargeable batteries or gasoline generators. The current is applied to the water by electrodes. The positive electrode, or anode, is usually mounted on a probe or boom and the negative cathode mounted to a metal boat hull or trailing cable. Fish caught in the electrical field orient towards the anode and swim towards it in a process called electrotaxis or forced swimming. When they get close they experience electronarcosis. As they lose equilibrium and roll over, biologists watch for the flash of their white bellies and net them. Some biologists use a cabled throwable anode with which they can draw fish to the boat from some distance away.

Electrofishing is highly size-selective; large fish are more vulnerable than small ones because they occupy a greater voltage gradient in the electrical field. In addition, they are more visible to dip-netters. Species with small or no scales (trout, catfish) tend to be more vulnerable than large-scaled species. Habitat selection can also influence susceptibility; fish that occupy surface waters are more likely to be captured than those in deep, open water.

Local water characteristics can influence the effectiveness of electrofishing. Pure water does not conduct electricity; rather, it is the ions dissolved in water that pass the current between the electrodes. Typically, impurities are present in sufficient concentrations for electrofishing, but some sterile headwater streams can be so clear electrofishing techniques do not work. Occasionally, biologists may temporarily add salt to some streams to raise water conductivity enough to allow electrofishing. At the other extreme, seawater is too conductive for effective electrofishing; the current disperses weakly in all directions rather than being concentrated between the electrodes. Water transparency is also critical because dip-netters must be able to see stunned fish. Fortunately, most waters in the GYE have excellent clarity except during spring runoff.

Electrofishing is exciting. A biologist can capture more large trout in a day of electrofishing than most of us can hope to catch in a lifetime of angling. It can be somewhat frustrating for a biologist who fishes to see just how many fish are out there compared to how few are caught on a fly rod. But, the effectiveness of electrofishing is tempered by the potential danger it poses both to fish and fishery workers. Workers must be insulated from the electrical current by waterproof rubber gloves and waders and must receive training in the proper and safe use of electrofishing as well as CPR. Fish can easily be injured or killed during electrofishing if

▲ PHOTO 7.5 Using a simple seining apparatus (top) allows fishery biologists to quickly and safely sample a small reach. This is a low-gradient, open stream in central Montana. (Robert Bramblett) This rainbow trout caught in gill net in Hauser Reservoir (bottom) can be released without harm if they are retrieved frequently and their struggling is minimized. (Justin Spinelli)

improper techniques or gear are used. Fish and wildlife agencies now require specific electrofishing equipment, power settings (direct current only), and procedures to minimize or prevent injuries to biologist and fish. Fish should be exposed to current as briefly as possible and should not be touched by an electrode. They must be allowed to recover fully before release to prevent predation.

The seine is perhaps the oldest and simplest fishing device. It consists of a rectangular fine-meshed net tied to poles at each end. Two netters pull the seine by the poles through the water, essentially sieving out any fish in it. Lengths of seines can range from a few meters up to about 30 meters; heights are generally a meter or two. Floats and weights are often affixed to the tops and bottoms, respectively, of longer seines to keep them open vertically. A pouch or "bag" is often located at the center of long seines to help contain captured fish. Seines tend to get caught on obstructions (logs, boulders) and are therefore most useful in open habitats such as beaches. They tend to be best for small, slow fish because large fish can swim fast enough to evade them. A common use of seines is to assess year-class strength of juvenile fish in lakes and reservoirs. They are also used to determine species occurrence in prairie streams just east of the GYE. Most streams and rivers in the GYE have too many obstructions for effective use of seines - especially after the fires of 1988 added large amounts of woody matter to waterways.

Gill nets superficially resemble seines with very large mesh openings. They are temporarily set in place and catch fish that swim into them of their own volition. They are therefore known as "passive" gears, unlike seines, which are "active" gears that are actively fished by netters. A gill net is basically a panel of coarse-meshed netting made of monofilament nylon fishing line. Those used in freshwater are typically 30 meters long and 1 meter deep. Lead weights (or a lead-core line) are attached along the bottom and floats (or a foam-core line) are attached along the top. Anchors and buoys are attached to the ends. When set in place, typically from a boat, the gill net resembles a long fence. Combinations of weights and floats can be adjusted to position nets along the bottom, surface, or at mid-water positions. Biologists sometimes set vertical gill nets (essentially horizontal gill nets turned 90°) that extend from the surface to the bottom to determine depth distributions of fish.

A fish that encounters a gill net may attempt to pass through it, but if the mesh size is appropriate for that size fish, it becomes ensnared behind its gills or fins. Mesh size is therefore critical to the success of gill-netting; small fish can simply swim through the meshes of a gill net and large fish cannot penetrate it sufficiently to become ensnared. Biologists can use this selectivity to target specific sizes, or they can use so-called

▲ PHOTO 7.6 This picket weir is set near the mouth of Red Rock Creek, Red Rock Lakes National Wildlife Refuge, Montana, looking upstream. Downstream migrating fish would be funneled into the trap in the center foreground whereas upstream migrants would be diverted to the trap on the right. The weir is made of closely spaced aluminum conduit and is held in place with metal fence posts. (Ryan Harnish)

"experimental" gill nets made of multiple panels of various size meshes to avoid size selectivity. Active, large fish that frequently move are more likely to be caught in gill nets simply because they are more likely to encounter the gear than less active fish or smaller fish that do not move as much. Nets are often set overnight to reduce avoidance by taking advantage of reduced visibility.

In the GYE, gill nets are commonly used in lakes and reservoirs to monitor fish abundances. Rivers and streams here tend to be too swift for effective use of gill nets, but they are commonly used in the lower Yellowstone River and other Great Plains rivers. Perhaps their most notable use here in the GYE is in the non-native lake trout suppression program in Yellowstone Lake, where up to 25 kilometers of gill nets are fished daily during summer. Gill nets are effective tools for catching fish, but need to be retrieved frequently to prevent injuries or death of fish as well as their needless struggling and suffering.

A variety of passive traps (hoop nets, fyke nets, minnow traps) can be used to collect fish. These may use bait to attract fish - a perforated can of cat food or salmon works well or, they may be designed to use "wings" of mesh to guide fish into them. All of them include hoops that support funnel-shaped mesh throats that guide fish into the traps and inhibit their exit. These gears tend to work best in reservoirs and lakes. The winged gears are placed in locations that fish tend to move through such as a narrow channel or entrance to a small bay.
Weirs are similar to winged trap nets, but are placed in streams and small rivers to intercept migrating fish. A typical installation involves a rigid fence of wickets or mesh positioned across the stream at an angle. Fish moving either up or downstream are channeled to live boxes with funnel-shaped openings. Weirs are very effective for assessing abundances of pre-spawn upstream-migrating adult trout and their downstream-migrating offspring. However, they tend to get clogged with drifting debris and therefore require frequent

▲ PHOTO 7.7 The rotary screw trap is a very effective long term sampling method. Fish are caught in a cone as the force of the water rotates it. This rotary trap is located on Skalkaho Creek, a tributary to the Bitterroot River in western Montana. (Al Zale, MSU)

▲ PHOTO 7.8 Snorkeling is fast, easy, and a fun way to count fish and inspect their habitat. These biologists are counting trout in Bridger Creek, Montana. (Christopher Guy)

cleaning and maintenance. They also can pose a navigation risk to canoeists. Yellowstone National Park operates a weir on Clear Creek on the eastern shore of Yellowstone Lake to monitor annual cutthroat trout spawner abundances. Sadly, whirling disease and lake trout have combined to reduce those counts to about 500 fish in recent years compared to over 70,000 in 1978.

A modern alternative to a weir for capturing downstream migrants is the rotary screw trap. It consists of a cone fitted with internal spiral planes mounted on a pontoon barge anchored in a stream. The open end of the cone faces upstream. The force of the water on the spiral planes causes the cone to rotate on its axis, much like an Archimedes screw. Downstream migrating fish that enter the open end of the cone are augured by the planes into a live box at the rear of the trap. A self-cleaning rotating-drum screen at the rear of the live box removes debris. Rotary screw traps are efficient and effective, but initial costs are much greater than weirs. Montana Fish, Wildlife and Parks operate a screw trap on Duck Creek just north of West Yellowstone (visible from Highway 191) to monitor migration of juvenile rainbow trout into Hebgen Reservoir.

Although relatively inefficient, sportfishing tackle is occasionally used in fisheries research, especially when only a few fish are needed and the trouble and expense of using a more conventional sampling gear are not worth it. Good examples are genetic or tissue sampling or collection of fish for a small radio-tracking study. Also, a fishing license is easier to get than a collecting permit, which involves considerable bureaucratic red tape. In some instances, the possible negative public perception associated with using conventional sampling gears (especially electrofishing) when people are angling can be avoided by using sportfishing gear. However, the public can also take offense when they see public employees angling while apparently at work.

Non-Capture Sampling Techniques

Fisheries biologists tend to enjoy capturing and handling fish and therefore gravitate to techniques that result in fish capture. Often, it is more efficient and just as effective to "sample" fish using less intrusive techniques that avoid capture and handling such as direct visual observation or hydroacoustics.

Snorkeling is an effective technique commonly used for counting fish in small and medium-sized, clear streams of the GYE. For example, it was used by NPS biologists to assess abundances of cutthroat trout in the Yellowstone River above Yellowstone Lake. A careful snorkeler can easily approach trout that would be alarmed by someone walking along the stream. Of course, some fish will be concealed, but conversion factors can be calculated to compensate for unseen fish. Fish can be identified to species and a trained observer can estimate lengths quite accurately despite the visual magnification property of water. Perhaps the biggest drawback to snorkeling in the GYE is water temperature; wetsuits and often drysuits are necessary to stay in the water long enough to complete counts. An alternative is to use underwater cameras and increasingly common technique in fisheries studies. Cameras mounted on remotely operated vehicles (ROV) towed behind boats can be used to count fish along transects in lakes and reservoirs. Cameras can also be lowered from boats to find fish congregations and identify spawning areas in lakes.

Trout spawning nests are known as redds. Females dig out depressions in gravel, usually at the upper ends of pools where clean, oxygenated water wells up through the substrate. After spawning, the females bury the eggs with more gravel from just upstream of the redd to hide them from predators. The disturbed gravel is often readily visible as a lighter area among surrounding darker substrates still covered with attached algae and fine sediment. Redds can be counted from the streambank and give a relative estimate of adult spawner abundances as well as information on spawning locations. Redds of larger females are more obvious, as are those of autumn spawners such as brook, brown, and bull trout (*Salvelinus confluentus*). Surrounding substrates are undisturbed by snowmelt runoff and have more algal growth in the autumn than in spring; water clarity is usually better in autumn as well.

Hydroacoustic technology such as sonar - SOund NAvigation and Ranging - uses transmitted sound to locate, count, and measure fish. Sound echoes off fish and is picked up by an underwater microphone. The intensity of the reflected sound correlates with fish size and the time lag between sound transmission and echo detection can be used to estimate distance (i.e., location). Hydroacoustics is useful when visibility is limited, such as in turbid or deep water. Applications can be "fixed" in one spot to record fish moving past it, or mobile, as when operated from a boat. Data are automatically recorded digitally and are later decoded manually or using automated computer programs. Hydroacoustics can collect vast amounts of data, especially in fixed applications, which can continuously record data almost indefinitely. A major disadvantage of hydroacoustics is its inability to identify fish to species, although inferences can usually be made based on size and location. Often, a limited amount of capture sampling to identify species is used to supplement hydroacoustics. Mobile hydroacoustics was used in Yellowstone Lake to document densities, sizes, and depth distributions of cutthroat and lake trout to better target their lake trout eradication gill-netting efforts and limit unwanted catch of cutthroat trout. Fixed beam hydroacoustics was used at Hauser Dam on the upper Missouri River by MSU researchers to estimate numbers of fish passed through the turbines and over the spillway.

Fish are tagged or marked to examine movements, growth rates, and abundances. A great variety of tags have been used in the past, but many of these were bulky and were found to affect fish survival, growth, health, and behavior. We now know that small, inconspicuous tags are best for the fish, but these can be easily overlooked. Some tags work best if noticed by the public. For example, in a study of harvest rate, an anchor tag must be external and conspicuous so the angler can report the catch. Anchor tags can be large, colorful, and imprinted with identification codes and return instructions. A T-shaped end is inserted with a grooved needle so that the T lodges behind a fish's bones to prevent tag loss. They are invasive to the fish and chronic inflammation often develops around the insertion point. An infection can affect a fish's growth and health. A less conspicuous and healthier tag is the visible implant tag - a tiny plastic chip imprinted with a unique alphanumeric code. It is inserted into the clear membrane behind a fish's eye with a hypodermic needle. Such a tag is readily visible, but only if you know to look for it. A similar method is to inject liquid colored rubber into the skin, which congeals into a permanent tattoo. Combinations of dots of different colors in different locations can be used to create an individual code for each fish.

Partial fin clips can be used when individual identification of each fish is not needed, such as for a simple mark-recapture abundance estimate. The tip of a fin is cut off with scissors and is recognizable until it grows back. Partial fin clips do not affect swimming ability. Complete fin clips do affect swimming and are therefore not used except sometimes for removal of the adipose fin, the small fleshy fin on the back just to the front of the tail. Full clips are used to trace salmon and trout to indicate hatchery origin.

Passive integrated transponder (PIT) tags are widely used in fisheries research. A PIT tag consists of a tiny computer chip and antenna encased in glass tube. No battery is involved. When activated by a PIT-tag reader antenna, the tag emits a unique alphanumerical radio signal. The tag is injected into the body cavity of a fish

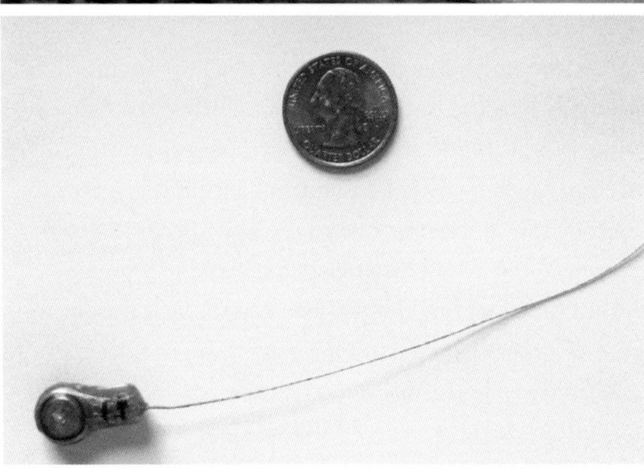

▲ PHOTO 7.9 A striped bass tagged with an anchor tag. The T-shaped internal end of the tag is lodged behind the interdigitated bones of the fish that extend down from the dorsal fin and up from the backbone (top). A visible implant tag is visible behind the eye (code "V15") in an Arctic grayling. Unlike some larger tags, these tags are harmless to the fish and does not change their behavior (middle). Miniaturization has made telemetry easier to track fish via satellite. A radio tag with trailing antenna is used for tracking small trout (bottom). The battery is at left and the tag's microcircuitry is visible to the right. (Al Zale, MSU)

with a hypodermic needle and can be read each time a fish is caught. Range of detection of the small (12 mm) PIT tags is limited to a few inches and therefore requires fish capture, but larger tags (23 mm) can be detected up to several feet away allowing for detection with mobile hand-held wand antennae or fixed antennae placed across streams. Cost is about three dollars per tag, which can allow tagging of large sample sizes, budgets permitting. MSU researchers are currently using PIT tags to determine movements of fluvial Arctic grayling in the Big Hole River.

As with land mammals, radio telemetry is used to determine the locations and movements of fish. We can know the habitats they occupy, when they move among them, and the routes they use. How habitat use and movements are affected by changes in the environment is especially informative. Migratory species, such as trout, can move great distances between spawning, summer, and winter habitats. Telemetry is an indispensable tool for documenting all of the habitats and mapping their location. Telemetry was used recently in the Yellowstone River above Yellowstone Lake to determine that most cutthroat trout there used the river for spawning and resided most of the year in the lake. Other uses of telemetry include assessing hybridization risk and entrainment of cutthroat trout into irrigation diversions in the Paradise Valley, determination of spawning locations of walleye (*Sander vitreus*) in Canyon Ferry Reservoir, and movements of rainbow trout relative to whirling disease infection risk in the Madison River.

Two types of telemetry transmitters (tags) are in common usage: radio and sonic. Radio tags are similar to those used in wildlife studies, but often use different radio frequencies that penetrate water better. They emit radio signals that are received by radio receivers tuned to the specific frequencies of the tags. The tags emit coded signals that allow biologists to differentiate specific fish. Directional antennas allow determination of a fish's location by triangulation. Range can be only a few hundred yards for small, weak transmitters or up to several miles with large, powerful tags. Tags can last from a few weeks to years and can be programmed to turn off and on as needed. Most of a tag's size is a function of its battery, which determines the tag's strength and longevity. However, large tags can affect fish behavior, survival, and growth, so biologists decide if the trade offs between signal strength and longevity and the data they collect are worth the potential harm to fish.

▲ PHOTOS 7.10 (left) Restoration of native fish populations is increasingly important for ecosystem management and to conform to the Endangered Species Act. Fishery biologists can isolate reaches by constructing artificial barriers to curtail upstream migration of nonnative species.
7.10 (right) A toxic chemical is introduced to reaches of streams to poison nonnative species so biologists can restore the native population. Although the chemical is destructive to the steam ecology, systems typically repair themselves in a short time. (Peter Brown)

A disadvantage of radio tags is that radio signals are rapidly attenuated by deep or salty water. They cannot be used for marine applications or in very deep lakes and reservoirs. Because radio antennas can be mounted on boats, vehicles, or aircraft, radio telemetry is useful for highly mobile species that require searching large areas.

Sonic transmitters emit coded high-frequency sound pulses that are detected by submerged hydrophones. Sonic telemetry works well in the large lakes and reservoirs of the GYE, but not in streams and rivers. Sonic tags require no obstructions like plants, islands, or river bends and are affected by turbulence like boat motors. The primary disadvantage of sonic telemetry is the requirement that the hydrophone be submerged, which precludes use from a vehicle or aircraft. Also, tags that are removed from the water, for example by a predatory animal or angler, cannot be detected. Radio tags implanted in fish are commonly found in or below osprey nests.

Both radio and sonic tags are typically implanted internally into the body cavities of anesthetized fish. An incision is made along the belly, the transmitter is inserted, and the incision is sutured closed. Tagged fish typically take a week or two to recover fully and return to normal behaviors. Biologists must spend a great deal of time locating tagged fish, especially if they move large distances rapidly and if signal strength is low; it is very easy to lose track of fish if their whereabouts are only infrequently monitored. Inexperienced biologists often underestimate just how much time and effort are required in a telemetry study. Tags can also be fairly expensive, around a $100 or more, which can limit sample sizes. Small samples limit the inferences that can be made about a population. However, the precision of the movement and location information that can be acquired by telemetry is much greater than with conventional tagging.

Anglers themselves can be sampled to determine fishing effort (i.e., angler hours or days spent fishing), fish catch and harvest estimates, and angler characteristics. On-site "creel surveys" are used if harvested fish must be examined by biologists. These surveys can be at access points where anglers are interviewed as they leave the fishery or "roving" surveys in which the creel agent moves through the fishery on foot or by boat and interviews anglers while they fish. Logistics usually dictate which is used. On-site surveys are expensive and time-consuming, which has led to development of off-site techniques such as mail, telephone, and internet surveys to collect economic, social, and effort data. Anglers are randomly selected from lists of license holders. Catch and harvest information from off-site surveys tends to be of low quality because of angler recall bias; they tend to better remember the bigger fish and better days. Moreover, species misidentification is common. However, off-site surveys can collect more information from more anglers (who are not hurrying to get home or still trying to fish) over a broader geographical area. Anglers sometimes tell a creel agent what they think the agent wants to hear when face-to-face in an on-site survey but anonymous surveys elicit more honest answers and is one of the major advantages of such surveys.

Fisheries biologists rarely kill large numbers of fish for research purposes. However, restoration of native fishes often requires eradication of introduced non-native populations that have outcompeted and displaced the natives. As the public and fishery managers learn to value natural fisheries and ecosystems, there will inevitably be more native restoration projects. In addition, fishery management under the Endangered Species Act may require eradication efforts.

Although repeated electrofishing can sometimes be used to successfully remove all non-natives in small streams, the use of fish toxicants is usually necessary. Rotenone, a plant derivative, and Antimycin, an antibiotic, are commonly used. They are applied using drip stations placed at intervals along a stream or from a boat in lakes. A natural (e.g., waterfall) or artificial barrier at the lower end of the treatment reach prevents reinvasion by non-natives from downstream. Single treatments are rarely fully successful because some young fish or eggs avoid coming in contact with the toxicant by occupying springs and seeps where clean groundwater enters. A second treatment in the following year kills these individuals before they get old enough to reproduce.

Both Rotenone and Antimycin hinder cellular use of oxygen and are taken up through a fish's gills. They are functionally harmless to terrestrial animals, including humans, unless inhaled, but Rotenone can kill aquatic invertebrates and amphibians. Fortunately, invertebrates usually recolonize streams quickly on their own.

Eradication programs can be controversial. On Cherry Creek, a tributary of the lower Madison River, Montana FWP teamed up with Turner Enterprise and the US Forest Service to establish a refuge for westslope cutthroat trout. Eradication of non-natives was held up for 4 years by administrative appeals and three lawsuits. Some segments of the public disagree with or see little use in native species recovery, especially if they use the existing fishery. Stressing ways in which restoration can improve angling, such as replacement of stunted brook trout fisheries by larger natives, can engender public support. Antimycin was used to eradicate brook trout in 1985 from Arnica Creek, a tributary to Yellowstone Lake. Other eradication efforts are currently ongoing or planned for several streams in Yellowstone National Park to restore fluvial populations of Arctic grayling and westslope and Yellowstone cutthroat trout.

Conclusion

The fishery resources of the GYE are extraordinary for their diversity, their health, and their public value. Unlike some other parts of the ecosystem, it is not readily appreciated because most of what goes on underwater is hidden from the public's view. Fisheries biologists use a wide variety of sampling and analysis techniques to collect information and gain inference on fish species distributions, abundances, sizes, growth rates, and ages in the GYE. These techniques all have inherent strengths and weaknesses that must be understood to collect unbiased data useful for managing the area's fisheries and maintaining their recreational, economic, and ecological values. The Yellowstone fishery is home to an outstanding recreation activity. It is also home to several important native species indigenous only to this region. Careful management based on good science will ensure this resource remains intact for many generations to come.

Chapter 8

If You Can't Measure It, You Can't Manage It: An Ecological Approach to Weed Management

Chapter 8

We are often given the impression that all weeds are bad. Dandelions in our lawn, bindweeds in our veggie patch, and spotted knapweed along our roadsides all need to be removed. But, is that true? And even it is it, can we realistically control all of them - particularly if our backyard is hundreds of thousands of acres of public land? This is the nature of the problem facing most public land managers in the West, and the Greater Yellowstone Ecosystem is no exception.

Virtually all public lands and national parks contain populations of non-native and invasive plants. For a multitude of reasons such species are considered undesirable and the general mandate is to remove them when they are found; a strategy known as early detection, rapid response. However, finding and controlling these species is difficult and expensive, particularly over large and rugged landscapes.

Bruce Maxwell and Lisa Rew were asked to develop and perform a survey of non-native plants in the Northern Range of the Park and develop a strategy for controlling non-native plant populations that occur over such a huge area. Is early detection and rapid response the most efficient and effective use of limited resources? Or might it be best to leave some of these populations alone?

Rew and Maxwell are not newcomers to these types of applied questions. Both have worked on weed management strategies for many years in agricultural and natural settings. They are also modelers. This means they use various statistical methods to build simulations of the real world. In turn, the simulations are used to develop a better understanding of the complexities of an invasion as well as the most efficient way to control them.

Here we see the approaches they have taken. Resource managers have used their probability models to search for previously undetected non-native plant populations, and to evaluate how changes in land-use may alter the presence of invasive species. Their most recent work evaluating when to manage suggests that sometimes, early detection and a passive response is the most logical management option.

J. Johnson

Lost in Montana (Trey Ratcliff)

Chapter 8

If You Can't Measure It, You Can't Manage It: An Ecological Approach to Weed Management

Bruce D. Maxwell and Lisa J. Rew

Bruce D. Maxwell, Department of Land Resources and Environmental Sciences, Leon Johnson Hall, Montana State University, Bozeman, MT 59715; Email: bmax@montana.edu

Lisa J. Rew, Department of Land Resources and Environmental Sciences, Leon Johnson Hall, Montana State University, Bozeman, MT 59715; Email: lrew@montana.edu

Lisa Rew has more information available at her website: http://landresources.montana.edu/rew

The overarching question we apply to our research work is: Can ecological information improve the ability of land agencies to manage the non-native plant species in their area? Stated another way: Will it pay off both economically and ecologically to spend some amount of time and money on understanding the ecology of these species in order to better target and prioritize their management?

Or, is it better to try to eradicate new arrivals when and where they can be found (early detection, rapid response - EDRR), and simply control more established species where they are known but not worry about looking for more of them?

To answer these overarching land management questions we need to address a series of questions relevant to plant ecology:

- How does one detect a plant population, particularly one in the early stages of invasion, i.e. when it is very rare?
- Is the population having a negative impact on the surrounding community, e.g. reducing the number, abundance, biomass and/or reproductive output of other species, or altering the amount of available water?
- If a population has established where is it likely to go next and how rapidly?
- How early in the invasion is early enough to make EDRR (i.e. eradication) work?
- When is it too late to stop the invasion because we cannot possibly target all the patches of the weed?

We will describe the approaches we used to answer these questions for the Northern Range of Yellowstone National Park.

The Problem with Non-Native Plant Species

Most of the plant species that have invaded Yellowstone National Park (YNP) are herbaceous flowering species that can be broadly divided into "monocots" (short for monocotyledons or one seed leaf) which are mainly grass species, and "dicots" (short for dicotyledons or two seed leaves; also called broadleafs or forbs). These species stop growing aboveground during the colder months, generally dying back to the ground each year but maintaining a root system from which they resprout the next year. Most of these non-native species occur in meadows mixed with native plants of similar size and general appearance, making them difficult to identify

▲ PHOTO 8.1 Public land managers fear non-indigenous plant species will drive out native species and replace them with a less healthy ecosystem. In fact, some non-native species are beneficial to native systems. (Jerry Johnson)

Non-Native Plants: Terms and Assumptions

The terms invasive, alien, exotic, non-native, non-indigenous and to a lesser extent weed are often used synonymously to refer to plant species not present at some previous point in history: pre-Columbian or pre-European human immigration. In reality there is little information on plant species occurrence prior to European settlement in the Greater Yellowstone Ecosystem. The first plant collections made in the Park were in 1871 by Robert Adams, a member of the Hayden Expedition. Frank Tweedy published the first flora list for the park in 1886. We tend not to use the term invasive species unless its invasiveness has been quantified, but use all the other terms interchangeably.

Management of non-indigenous plant species (NIS) within a rangeland or wildland ecosystem is typically based on the premise that they are invasive and therefore, must be managed. Noxious weeds are a subset of NIS legally designated by federal and individual state governments as having negative impacts upon crops, livestock, and native plant communities. However, the selection criteria are not consistent nor generally quantitative.

In reality any plant species that persists in a community, either native or non-native, has the potential to be invasive given optimum environmental condition. The key here is 'optimal' conditions: population growth rates may vary due to habitat suitability and can vary across time as resource availability fluctuates. Invasiveness can be defined as a population increasing in density and/or spatial extent.

The result of the assumption of invasiveness for all non-native species is that the same level of management effort is usually applied to all populations of the species, regardless of the environments in which they grow, even though population growth rates (i.e., the level of invasiveness) vary across the landscape. In addition, the impacts that a species' population is having on the environment surrounding them will also vary. Impacts include diminished native plant diversity, alteration of community structure and composition, threats to rare and endangered species, reduction in wildlife habitat and forage, alteration of disturbance regimes, depletion of soil moisture, changes in nutrient dynamics, and changes in the structure and function of belowground communities. However, few studies have evaluated populations in a range of different environments.

It seems intuitively unlikely that the same species has the same impact under all conditions when we know that the degree of invasiveness varies. The invasiveness and impact of a non-native population is not merely a function of it's presence. It is also the species' biological attributes, site-specific environmental conditions, and the composition of the surrounding plant community. To assume that all populations are capable of the same degree of invasion and impact, and that the impact is consistent across a varied landscape, may result in a misdiagnosis of the invasion and impact potential of the species. This mistake may result in inappropriate or inefficient management.

without a trained eye and often a taxonomic key. A total of 187 non-native plant species have been recorded within the Park, which comprises 15% of the total plant species. Many of these species have very pretty flowers and one may argue, add to the aesthetic value of the ecosystem. It is also the case that some of them, such as timothy grass (*Phleum pratense*), can have higher levels of nutrients in them than the native grasses and therefore are useful as a food source for elk, bison, and other grazing animals.

As with most issues there is a whole gamut of views. At one end are those that argue that all ecosystems or habitats change over time so the presence of non-native plant species is not a problem. Others would argue that these species have the potential to destroy the ecosystem by displacing native plants that support the food web that connects everything from bacteria to grizzly bears. At the extreme end of the range are those who consider the sheer presence of a non-native plant to be an unacceptable impact. Rather than just choose where to stand along this "no-problem to big-problem" continuum we are collecting data and have developed a framework to help us and land managers determine how to address this problem in a more rational way.

Quantitative information collected to determine the impact of non-native plant species indicates that there is great variation in the impacts caused by invasive species and even between different patches of the same species. That is, in one patch a species might have a negative impact by displacing native species. In another, the impact may be negligible or non-existent. Ecologists and land managers almost always default to the *precautionary principle* and so would rather assume that the introduced species is likely to have some detrimental impact and thus would recommend removal just to be safe.

This reasoning seems logical from a conservation point of view, but two points make the issue more

▲ PHOTO 8.2 Elk grazing in an area of the Northern Range of Yellowstone National Park that is highly infested with the non-native grass timothy (*Phleum pratense*) and other nutritious "weeds". The grass was introduced by the National Park Service as forage for bison and elk. Today, it is found throughout public and private lands in the region and is an important food source across the Greater Yellowstone. (U. S. Fish and Wildlife Service)

questionable. First, it is costly to detect and remove non-native plant species. Second, the methods used to remove these species, primarily herbicides, can have extensive off-target effects - often killing the native plants that may offer the greatest competition to the invaders. The reason for these off-target effects is that most herbicides, especially those used for general weed management in range and wildland areas are not selective to a specific species. A particular herbicide targets either grasses or broadleaf species. If one applies a broadleaf herbicide to control, for instance, spotted knapweed (*Centaurea stoebe*, formerly *C. maculosa*) it will also kill the native broadleaf species present in the same area such as sticky geranium (*Geranium viscosissimum*), larkspur (*Delphinium spp.*), etc. The result of killing off the native species along with the non-natives is that there will be bare ground left behind. As non-native species tend to be adapted to disturbed environments, it is more likely that the same or other non-native species will re-establish in these areas, at least initially. Another consideration is that unless the herbicides are applied at an optimal time - when the plants are actively growing and with all the water and nutrients they need - the success of the herbicide is far from 100%. In fact, under many field conditions killing 80% of the target plants would be considered a success. This represents an economic waste as well as an inefficient control method. Although it must be said that the timing of other control methods, such as pulling by hand, is even more time critical and often less effective and, biological control methods take a longer time to reduce the target population. Herbicides are a useful tool but due to the inherent problems mentioned above they should be used to target the most rapidly increasing populations. This includes those acting as sources (i.e. sending out seeds to start new populations across the landscape) and those having the most negative impact on the surrounding community.

Field Detection and Predictive Maps

Detecting invasive plant populations is difficult and tedious because in areas such as the Greater Yellowstone Ecosystem they are essentially rare, even though they can be locally abundant. Someday it may be possible to reliably detect non-native invading plant species with remote sensing, but for now it is of limited help in the detection process due mainly to the scale of the imagery, the size of plant patches, and cost. Instead, we rely on ground based field methods to maximize the efficiency of the search for the metaphorical needle in the haystack. Sampling methods have been designed which capitalize on our knowledge of how species were introduced to an area and how they disperse.

The occurrence of non-native plant species in Yellowstone, like other places, is associated with humans. Some species have been introduced as flowers to decorate the lodges and their dining rooms (e.g. Dalmatian toadflax, *Linaria dalmatica*) and for other ornamental purposes such as lawns (e.g. Kentucky bluegrass, *Poa pratensis*). Others, such as timothy grass, were introduced to increase the forage for wildlife and domestic stock animals back in a time when Park management was dictated more by utility than conservation. A third type of introduction is more accidental and is along roadways where the sand, gravel, and rock material used to build the roads or provide for increased traction on ice in the winter, comes from stockpiles outside of the Park that are infested with non-native plant seeds. The Park Service along with other agencies recently introduced a weed-free certification for gravel pits to reduce this introduction vector. Other roadside introduction is related to vehicles themselves, with seeds and other plant parts being moved around on vehicles. This third type of introduction relates to large areas of the Park and should be considered when identifying the likely places to find new populations.

During 2000 our team of plant ecology scientists at Montana State University was asked by the Park Service to create an inventory of the non-native plant species and their distribution in the Northern Range of the Greater Yellowstone Ecosystem. Given that the area of the Northern Range is so large, 152,785 ha, and the initial budget was small and finite, it was abundantly clear that we could not sample the entire area. We used the knowledge that non-native plants tend to be associated with roads and other rights of way and decided to start sampling from those areas and move along a perpendicular transect away from them. In this way, we

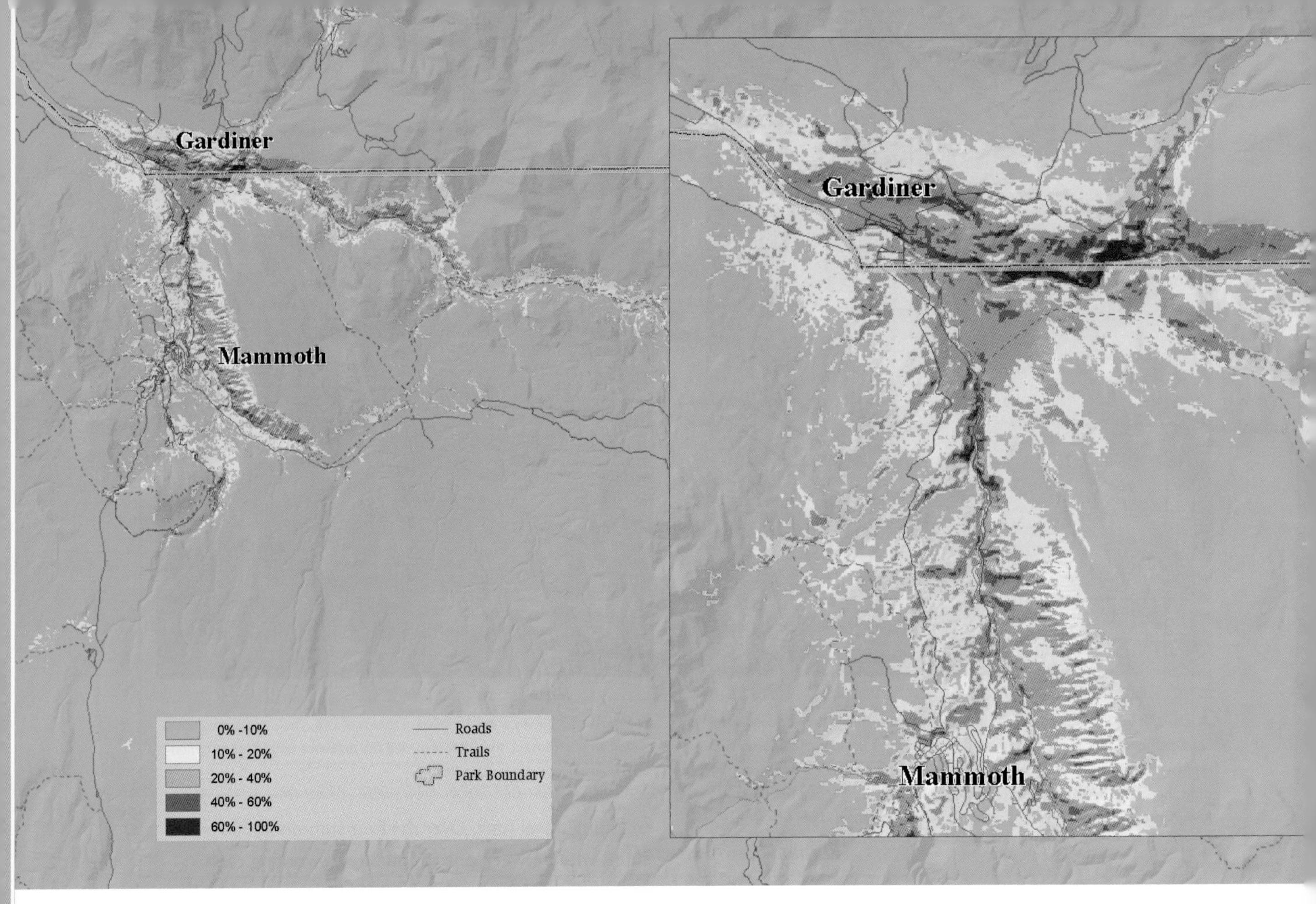

▲ FIGURE 8.4 Probability of Occurrence maps for Dalmatian toadflax *(Linaria dalmatica)* in the Northern Range of Yellowstone National Park. The decline in toadflax occurrence with distance from roads and trails, the preference for south and southwestern aspects, slopes between 15 and 30 degrees and other environmental variables are demonstrated by these map. The maps are useful tools for the park service to use to target the likelihood of future infestations. (Invasive Weed Ecology Lab, MSU)

any large area because any way to help land resource managers prioritize which areas to search for new weed populations will increase the efficiency and effectiveness of management.

Once species have been found or their location predicted with PO maps the next stage is monitoring for invasiveness and impact of populations. We have found that the PO values are a reliable surrogate of habitat suitability for a species. As the habitat becomes more suitable, the plant is more invasive. For example, Dalmatian toadflax prefers a more southerly aspect on a 15-30 degree slope. In these locations, it will tend to be increasing in density and area and therefore is likely to produce more seeds. If those seeds fall in a more suitable habitat, they have the potential to produce more new populations. Using the PO values, the manager responsible for controlling non-native species can prioritize species and patches for management based on whether or not they are in areas of high or low PO. High PO would receive management first.

Determining the Rate of Spread

In 2007 and 2008 we began to retrace some transects from 2001-2004 to determine what changes occurred to the populations of non-native plant species. Using the GPS, we were able to retrace our tracks along each transect to determine if there were any new patches, if any patches had gone extinct, if they had grown or, if the populations are unchanged.

Using the presence and absence data from the same transects observed in two different years we are able to estimate the probabilities of colonization and extinction as well those areas remaining unoccupied or occupied. The probabilities of changes in a population over time are called Markov transitions. We found, once again, that PO values, as a surrogate for habitat suitability, was a significant covariate or driver of these transitions. That is, PO value is a good positive predictor of where new patches are likely to occur; the higher the PO value, the higher the probability of seeing a new patch of the target species. Similarly, low PO value is a good predictor of where patches are likely to go extinct without intervention. We also found that distance to the nearest existing patch along a transect was a robust predictor of new colonization, this phenomena also increased along the PO gradient. Equipped with these different transition probabilities we can develop new maps detailing different rates of spread over the landscape, and over time. This ability to simulate an invasion of a non-native plant species also allows for the simulation and evaluation of different management strategies for curtailing the invasion.

▲ PHOTO 8.4 Management of non-indigenous plant species with herbicides is the most common form of management and is typically conducted close to roads. Unfortunately, unless the chemical treatment is applied at the correct time and in the correct way, it can be relatively ineffective. Our models show that sometimes such treatment is not warranted because the infestations are so pervasive; however, given early detection of new outbreaks herbicide application may be cost effective. (Erik Lehnhoff)

Simulation of the Manage or Not Manage Dilemma

Based on the ecological knowledge that we have gained about how plant invasions proceed including the variability among species, and populations within each species, we can return to our question: Is the scientific information valuable enough to improve the management of these invasions or would it be better to just put all of the effort into eradicating or controlling populations when found? Ideally, we would have ten or more seasons of data to track changes in the populations on the ground before making such recommendations; during that time some patches would spread and give rise to new patches. However, public land managers rarely have the time to wait so long to make a decision. As an alternative to long term field monitoring we created two theoretical simulation models based on the transition data we had already collected.

When using theoretical models it is considered good technique to use competing models to ask the same question; that is, models with difference in their structure for simulating the same phenomena. If both produce the same response/results one has greater confidence in the results. Our first model assumed a highly simplified landscape with no variability in habitat quality but three classes of populations: extinction, equilibrium and source class, and four management strategies. The second model assumed a more variable background that provided a range of habitat suitabilities similar to the PO gradient for establishment and invasion potential.

The simulation experiments conducted with the first model (constant habitat suitability) revealed some interesting outcomes. First, the Early Detection Rapid Response (EDRR) management approach was more successful than the other management strategies at reducing the number of populations in the first seven years. The constraints of this approach were that detection and control had to successfully remove more than 67% of the populations and, the managers had to randomly choose the populations to manage regardless of their location relative to the road. These requirements for success with the EDRR strategy are stringent and may be very difficult to achieve. A detection and removal rate of 67% of all populations would be very optimistic for most of the species that have invaded YNP and the GYE. Further, eradication far from existing roads or trails is time consuming and expensive.

Monitoring to identify source populations, followed by their management (Monitor 50% and Manage 50% time), became the best management approach for reducing the rate of invasion after approximately nine years. This is true even when only five populations could be managed in any given year. Thus, this model suggests that ecological information that can help identify source populations is of value and may be the best approach for managing invading populations. The long and short of it is, *if* detection and control rates are high and imposed as soon as a species arrives in the area, EDRR can work well, *but*, as soon as a species has been present for around ten or more years there are too many populations for chemical control to work if it is just applied to random populations or those near roads. After around ten years, knowing which populations are being invasive and

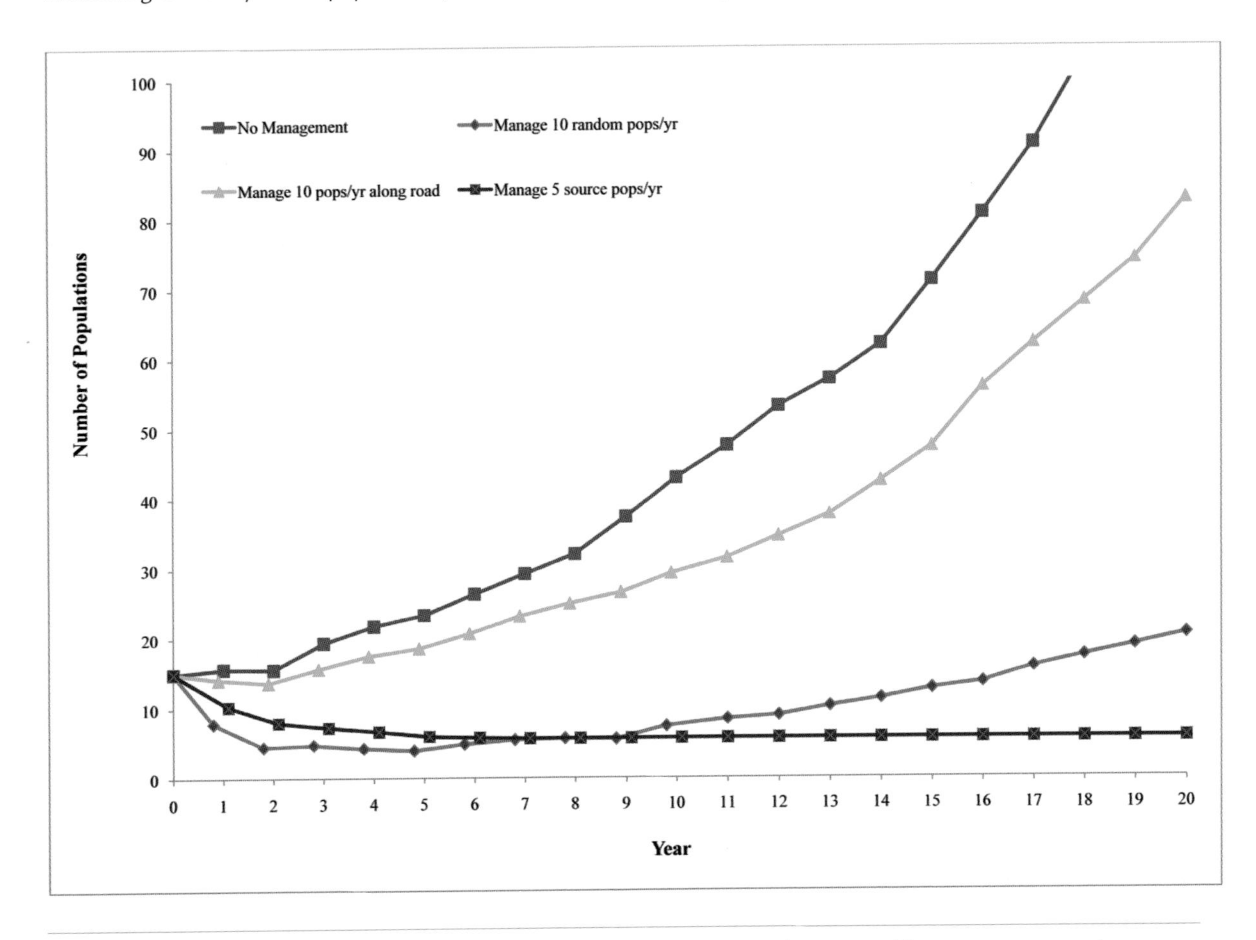

▲ FIGURE 8.5 Based on our simulations and incidence of non-indigenous plant population, we developed responses to different management strategies over a 20-year period. We determined that managing new source populations is the best management approach. If new infestations are eradicated early then control is effective. This approach depends on early detection in remote locations using models based on field data and real world observations. (Invasive Weed Ecology Lab, MSU)

spreading seeds to other areas will provide better overall reduction in the number of populations. For most non-native species in YNP and the GYE, this latter scenario is the most realistic.

The second simulation model – with variable background environment –shows a similar pattern. EDRR will work if it can be implemented early enough in the invasion process so that a large proportion of populations can be detected and eradicated without regard for a population's potential for being a source for new patches. However, if one is slower to locate new species and their populations, gaining ecological information about the species (monitoring) and using that information to target (kill) source populations will be the superior management approach.

These are results from theoretical simulation models, albeit they are based on some field data. Theoretical models still make many people nervous, as all models are based on assumptions that cannot reflect the full complexity of the natural world. Models have the advantage of helping us test assumptions and determine driving processes that would take decades of fieldwork to ascertain. Therefore, we will continue to collect data for the next few years using the same transects and will continue to improve and modify the second model with more empirically based information. Over time, we will be able to suggest best management strategies with more certainty.

Conclusion

Returning to the original and overarching question: *Can ecological information improve the ability of the land agencies to manage the non-native plant species in their area?* The answer from our fieldwork and modeling is "yes". The occurrence of non-native plant species varies over the landscape and there is no evidence that any of these species will "take over" and form a monoculture over the entire area. There are parts of the landscape that are more or less suitable for a particular species due to differences in environmental variables such as aspect, slope, soils and other vegetation. There is also a trend of more non-native plant species being present in disturbed areas; this is demonstrated by higher frequencies of non-native species close to roads and trails in our example. The correlation with disturbance is very important, whether the disturbance is more anthropogenic such as road maintenance or the addition of gravel, or natural as in the case of wildfires and floods. As a general rule, an area will be more susceptible to invasion by non-native plants if it is disturbed. This needs to be considered when thinking about the likely benefits of control. Our sampling and simulation models demonstrate that most non-native species in the GYE are already too frequent for Early Detection and Rapid Response to have a positive effect of reducing population numbers. It may be possible to use EDRR to reduce numbers in a specific small area such as 10 ha but likely, there are so many populations in the surrounding area that new populations would easily reinvade.

There is a bright side to this dark cloud for public lands managers: *if* we spend some time detecting and monitoring (i.e. measuring) a small number of weed populations, *then* we can extend this information to an entire area of interest. We can prioritize the species and populations to manage and can thereby reduce the total number of source populations within an area such as the GYE. As non-native plant species are present on most public lands and national parks, this prioritization approach will work there too. In addition, park and resource managers have many demands on their time and budget and no state or federal agencies have enough money to manage all noxious weeds in their area, let alone all the non-native species. Using the information described would change the way non-native species are currently managed and could prove to be an effective economical and ecological way forward.

ACKNOWLEDGMENTS: Parts of the work described were funded by the National Park Service Yellowstone Inventory and Monitoring Program, the Center for Invasive Plant Management and the USDA Forest Service. We would also like to thank the Weed Ecology group at Montana State University, particularly Erik Lehnhoff and Tanya Skurski.

Chapter 9

Yellowstone Extremeophiles: The Life of Heat-Loving Microbes

Chapter 9

The hot springs of Yellowstone are fatal to several unfortunate visitors every year. The hottest springs are close to boiling and some are as caustic as battery acid. Yet, even in these extreme conditions life thrives. The rainbows of colors seen in the park's thermal features are thermophilic (heat loving) microbes and represent a level of biodiversity unrealized until just a few years ago. The multicolored slimes flowing along the edge of thermal water are actually ecosystems of microbes.

Mark Young, along with others at the MSU Thermobiology Institute (TBI), investigates the myriad of life forms in the microbial mats and open water of thermal features; they have found life virtually everywhere they look. Mark's main interest is viruses - the most abundant biological units on the planet and also among the smallest. Working with samples from the park and a powerful electron microscope, Mark has identified elaborate structures in viruses with an evolutionally history extending back more than 3 billion years. One of his goals is to figure out how to use these intricate virus lattices to deliver drugs to humans. These "virus cages" could act like tiny Trojan horses, carrying therapeutic drugs directly to the cells that need them while minimizing unintended damage to healthy cells. The fact that they survive in extreme environments like thermal features means they are robust vehicles.

One of the unique features of the TBI group is their emphasis on teaching others about the thermobiology of the Park. They have produced educational videos, books, and courses for the general public so they can share what they are learning about the unseen life in Yellowstone. Discoveries in the bacterial mats promise to continue to be exciting and may one day unlock novel medical treatments or help solve our energy dependence. The thermophiles will continue to teach us the complexity of Yellowstone.

J. Johnson

Norris Geyser Basin (Trey Ratcliff)

Chapter 9

Yellowstone Extremeophiles: The Life of Heat-Loving Microbes

Mark Young and Jennifer Fulton

Mark Young, Department of Plant Sciences, 307 Plant Bioscience Building, Montana State University, Bozeman, MT 59717; Email: myoung@montana.edu

Jennifer Fulton, Ph.D. Graduate Student, Department of Plant Sciences, 307 Plant Bioscience Building, Montana State University, Bozeman, MT 59717; Email: jennifer.fulton@msu.montana.edu

Check out the many outreach and learning opportunities at the TBI website: http://tbi.montana.edu

The crime was hideous, a double murder late in the night on the boardwalks surrounding Old Faithful. The perpetrator seems to have disappeared without a trace. But not quite. The perpetrator's bloody handprint remains in the mud by the victim's side. Today, as any fan of crime mystery television can tell you, DNA sequencing can quickly and accurately identify the owner of any microscopic pieces of evidence left behind.

The handprint alone is enough to identify the perpetrator using methods that had their origins in discoveries made in Yellowstone years before. What most people do not know is that the ability to sequence tiny amounts of DNA is due to the discovery of a bacteria that grows in the hot springs in Yellowstone National Park and other thermal areas around the world.

In the 1960s, microbiologist Thomas Brock was studying microbes that grow in Yellowstone's hot springs. He isolated a novel bacterial species, *Thermus aquaticus*, from the Great Fountain area of the Lower Geyser Basin. This organism was thriving at 70°C (158°F). This meant that all of the enzymatic machinery inside this single cell organism was functioning at a temperature much higher than that of any other known organisms. *Thermus aquaticus* has to copy its own genetic information with an enzyme called a DNA polymerase in order to survive and replicate. If *Thermus aquaticus* lives at 70°C, its DNA polymerase must function at that temperature as well.

The discovery of this DNA polymerase enzyme and its application in a process called *Polymerase Chain Reaction* (PCR for short) resulted in a Noble Prize and a multi-billion dollar industry. PCR essentially operates as a molecular photocopy machine, allowing for the copying and amplification DNA from a very small starting amount. PCR has become a routine tool in laboratories all over the world. It is the basis for our molecular understanding of the relationships between all forms of life, it allowed scientists to sequence the DNA genome of any organism, including the complete human genome, it allows the medical world to better understand and design treatments for genetic diseases, and it lets the CSI technician ID the culprit in our crime scene.

Upper Temperature Limits of Life

Microbiology has long focused on organisms with medical relevance. Microbes that can infect us and other vertebrates thrive at the temperatures in which the host organism lives; 37°C for humans. Few scientists thought that life could exist in the extremes of temperature and other parameters, that exist on the planet. Extreme heat in Yellowstone or cold in Antarctica, high salt concentrations as in the Dead Sea, heavy metals concentrations found in abandoned mines, among other stresses, were all thought to inhibit life. However, a closer look at these environments reveals that they not only support life, many organisms actually thrive there. We now know that life exists at temperatures up to and exceeding 121°C. This is remarkable, considering that 121°C is considered the "gold standard" of sterilization and is used to kill all previously known microbes. Most organisms are literally cooked at these temperatures, with their proteins denaturing, much like a cooked egg; lipids and other molecules becoming unstable and melt. Understanding how this organism thrives at such extremes can help us understand the fundamental nature of these biological molecules and may even help us understand how life first began on earth.

A Short Explanation of Microbial Life

The word microbe is general term and simply refers to the small size of an organism. If you need a microscope to see an organism it's likely a microbe. Some microbes are pathogenic; *Bacillus anthracis*, causes anthrax, *Yersinia pesitis*, the causative agent of bubonic plague (the Black Death), are but two famous examples. Other, benign, microbes that surround us, cause no harm, and often aid us. There are more than 2000 commensal (beneficial) microbes in and on our bodies, while there are only approximately 100 known pathogens. Few of these pathogens are in residence in us at any given time. Those thatare present are being kept in check by the commensals. In our gut, they help metabolize food and glean all the available nutrients that we could not access without them, they synthesize vitamins that we cannot make or get from our normal food sources. For example, the human gut cannot efficiently extract nutrients from complex carbohydrates (starches). The end products of human digestion of these starches become the energy source for colonic microbiota like *Clostridium and Firmicutes* species which then provide as much as 15% of the human energy requirements. These beneficial bugs also keep pathogens at bay simply by outcompeting them for space and nutrients.

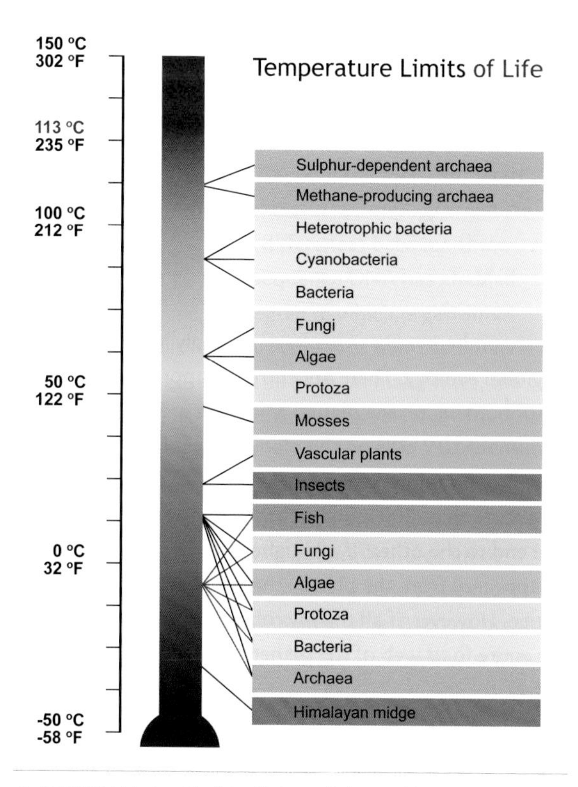

▲ FIGURE 9.1 Most multicellular life forms (Eukaryotes) live at temperatures below 113° F (45 C). Hyperthermophiles, single celled microbes, live at temperatures greater than 176 °F (>80° C). The highest temperature known than life can survive is 266° F (130° C). Thermophiles thrive between 113 -175° F (45-80 C). (Thermal Biology Institute, MSU)

DNA Analysis and PCR

The ability to read the genetic code of an organism has become indispensable for biological research. By reading the sequence of the four nucleotide - adenine (A) , guanine (G), cytosine (C), and thymine (T), that make up a molecule of DNA (Deoxyribonucleic acid), scientists can read the genetic instructions used in the development of all living organisms. Knowing how to decipher the DNA information is a central theme of the life sciences and impacts the understanding of evolutionary biology, the diagnosis of disease, and the law, to name a few.

The ability to sequence DNA is significantly enhanced by a process called PCR (polymerase chain reaction). PCR is used to amplify or replicate a single or few copies of a piece of DNA. In many ways PCR is like a molecular photocopy machine that allows for the specific amplification of small amounts of DNA. PRC allows researchers to generate thousands to millions of copies of a particular DNA sequence. This breakthrough has multiple uses. If doctors can identify the DNA segment that provides for immunity to a disease, they can replicate the strand and use it to treat patients, food crops can be genetically manipulated to select for desirable traits, small tissue samples can be traced to help solve crimes. Mark Young and his crew use PCR analysis to discover new life forms in Yellowstone hot springs.

PCR requires a small sample, often less than a drop, from which the DNA is amplified for sequencing. During the PCR amplification process, each type of nucleotide (A, C, G, or T) is chemically marked with its own unique fluorescent tag. The sequence of the amplified DNA is determined from the labeled DNA when it passes through a sizing matrix that can separate DNA molecules that differ by only one nucleotide. As the DNA molecules in the sizing matrix pass by a laser, a detector records the color of the fluorescent tag. A computer is then used to assemble the sequence of the DNA analogous to determining the order of four different colored beads on a long string of beads. The order of the long string of DNA is the genome of the organism and it can be millions to billion of nucleotides long. From there, specific gene sequences can be studied and tested for composition and function. In the case of thermophiles and hyperthermophiles there are genes allow the organism to make a living in high temperatures, allow them to exist in extremes of pH (potential of hydrogen) or low oxygen environments.

▲ PHOTO 9.2 The bright colors seen in Yellowstone hot springs are often formed by microbial communities that form mat-like structures. The color of the mat can often be used to identify the organisms and the temperature at which they are growing. The green colored mat is made up of a photosynthetic organism that grows at a lower temperature than the bacteria that makes up the red colored mat. (Thermal Biology Institute, MSU)

How Do These Organisms Survive Temperatures That Would Kill Most Life?

This question has been one of the fundamental questions driving research in Yellowstone hot springs. Answering this question would not only provide insights into these unusual organisms, but also how life may have first evolved, as well as give us biotechnology tools for working with biological systems in these extremes.

Unlike warm blooded animals, microbes have no ability for internal thermo-regulation in the face of non-optimal external conditions. Instead, all of their internal cellular "machinery" must adapt to the temperature

▲ PHOTO 9.3 At Mammoth Hot Springs the terraces are built from the precipitation of carbonite. Different sizes and shapes of precipitates form depending on the particular combination of chemical, biological and physical factors coming into play in any given micro-zone of the terraces. This complex relationship creates one of the most iconic thermal features in the Park. (Thermal Biology Institute, MSU)

of their environment. Some of the adaptations we see in thermophiles include unusual lipids (fats) not found in mesophilic (moderate temperature loving) organisms. Most of these organisms have very efficient DNA repair systems and proteins that coat or coil the DNA to protect it from an environment that is inherently damaging to genetic material. The proteins produced by these organisms often fold into conformations that are inherently stable and thus resist denaturation at high temperatures. However, the bottom line is that we still know relatively little about many of these heat adaptations.

How We Do Our Work

For centuries microbiologists have studied organisms that grow in medias cooked up in the laboratory. This is true for both medically relevant and environmental organisms. This has served us well. It allows us to grow a single organism to high concentration and study it in detail. Experiments are reproducible, and we can often manipulate these organisms *in vitro* to produce proteins, antibiotics and even genetic material that have numerous beneficial applications in medicine and biotechnology. Petri dishes of bacteria are familiar to everyone who has taken high school biology.

However, there are certainly enormous numbers of organisms that simply do not grow well in these conditions. Altering media to expand what we can grow has worked well to date. But, we are still faced with a

Microscopy

While molecular tools have come to prominence in the last few decades, the microscope remains a fixture in most microbiology labs. The microscope has been a fundamental tool in understanding microbes since the birth of microbiology more than 300 years ago. Huge advances have been made since Antony van Leeuwenhoek's first design. Today we have powerful electron microscopes capable of viewing single viruses and strands of nucleic acid at 100,000x amplification. Improvements in the light microscope and the use of fluorescent dyes or antibody-linked dyes have resulted in incredibly powerful tools. Observation of unstained cells can tell us about numbers of cells in a culture or the morphology of the cells. Observation of differentially stained or immuno-flourescently labeled organisms can provide information about composition of the cell wall, whether the cell is alive or dead, the number of chromosomes present and even the localization of specific proteins within the cell.

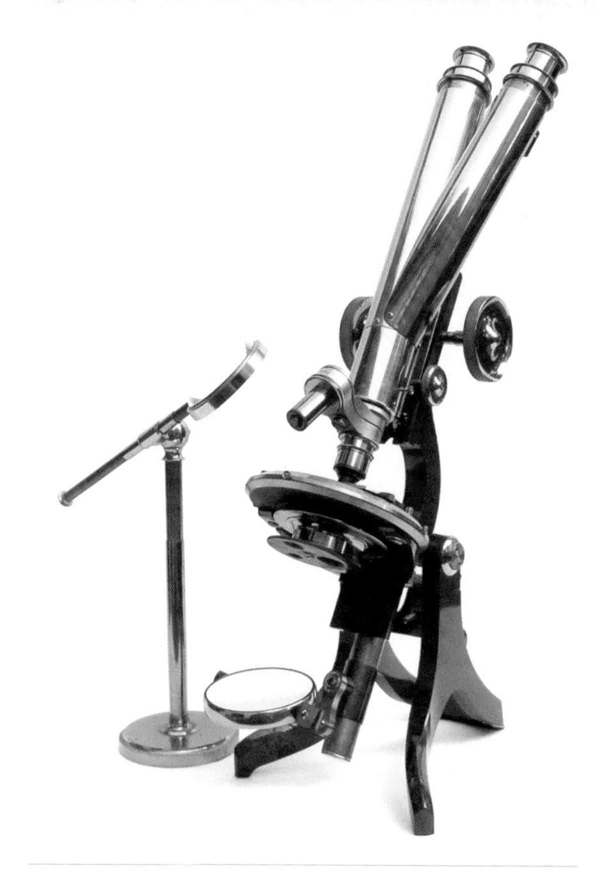

▲ PHOTO 9.4 The first light microscopes were made in the late 1590s which allowed some of the first single celled microbes to be seen. Modern light microscopes are indispensible for observing the microbes that live in Yellowstone hot springs. (Allan Wissner)

vast level of diversity in the environment that doesn't match what we have in culture. Furthermore, in many instances, the organisms that we are able to consistently culture out of an environmental sample are not the major players in the environment. *Sulfolobus* species are one such example. Various species and strains of *Sulfolobus* have been grown in enrichment cultures out of hot acidic pools all over the world. In fact, if a pool is between pH 2-4 with a temperature between 70-80°C, some species of *Sulfolobus* is likely to grow if the right media is present. One result will be the rotten egg smell ubiquitous at thermal sites. If we study these environments,by other techniques we find that *Sulfolobus* species often make up only a small fraction of the inhabitants.

To understand better who all the players are in these environments scientists have been recently using culture-independent techniques. What this means is that we spend a great deal of time in the thermal regions of Yellowstone sampling hot springs, mud pots, and the other thermal features. When we return these samples to the lab, we can sequence specific genes, aided once again by the *Thermus aquaticus* DNA polymerase, and we find organisms that we would never detect by trying to grow them in the laboratory. What we discovered when we started to look at pools in Yellowstone using these methods, was that the diversity was astounding. The vast majority of the organisms were ones that had not been previously identified.

In just the past few years sequencing technology has once again made huge leaps forward. It is now possible to sequence entire communities of organisms at once. The speed and ease of sequencing has created a situation in which we can quickly obtain more information than we are currently able to process. We can see that these environments are full of novel life forms and viruses that are going about their business in ways that we don't yet fully understand.

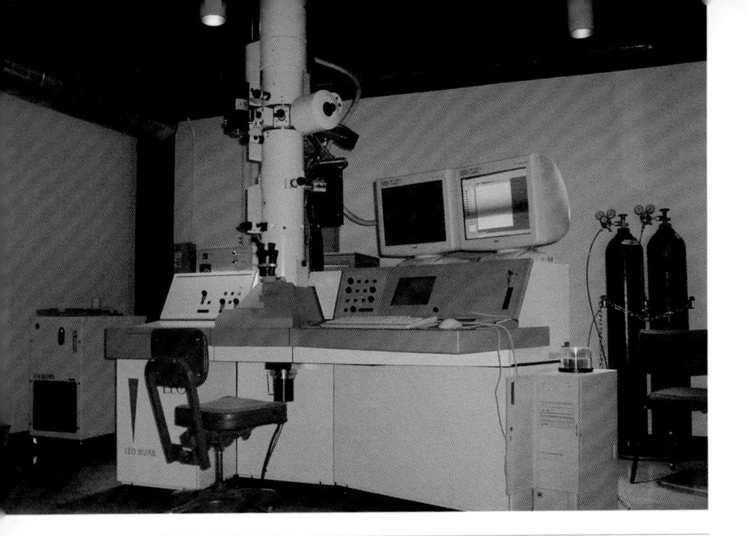

▲ PHOTO 9.5 A modern electron microscope allows observing details of cellular structure at much higher resolution than is possible with a light microscope. Using an electron microscope to look at microbes from Yellowstone hot springs allows new insights into the cellular structures that allow life to exist at high temperatures. (Thermal Biology Institute, MSU)

▲ PHOTO 9.6 Scientists carefully sample Yellowstone hot springs. Ongoing research in Yellowstone's hot springs is making fundamental scientific discoveries about the diversity of life, our understating of life at high temperatures, and about the evolution of life on earth. (Thermal Biology Institute, MSU)

A Day in the Field

Scientists are increasingly interested in studying these microbes and viruses in their natural environment, essentially using the hot springs themselves as the incubators and culture flasks. This work obviously requires new skills not usually associated with bench science. Yellowstone National Park is a dynamic environment for both the researchers and the organisms we are studying. Some studies are looking at changes in microbes over time, requiring monthly or weekly sampling points. In the summer, researchers spend long days hiking in the backcountry of Yellowstone locating hot springs and sampling over several time periods. The backcountry of Yellowstone is not only inhabited by a number of intimidating macrofauna, like bison and bears, requiring scientists to travel in groups of at least two, but the terrain itself can be quite treacherous. More than one graduate student has been initiated into field-work by stepping through the thin crust around a boiling acid hot spring.

Winter sampling requires long days outside in frigid conditions, often below zero (Fahrenheit). Travel into the backcountry might require snowshoes or skis. Working around boiling hot springs provides an unexpected challenge in the winter as well. While the boiling springs provides a relatively mild environment for collecting the sample itself, upon departing for the long hike back to the road, one discovers that all their warm winter layers are soaked through from the steam. Hypothermia in such an environment is a serious risk.

Extracting water samples from boiling acid often requires special tools to avoid falling into the pool. Scientists suspend sampling bottles and measuring instruments like pH meters and temperature probes from telescoping poles in order sample from the relative safety of the more stable ground away from the pool's edge. Often samples must be collected in as sterile a manner as possible in order to avoid contamination from surrounding soils or the researcher herself. This is no small challenge in the backcountry. Pre-sterilized vials and on-site filter sterilization along with working as close to the hot springs as possible, helps avoid contaminants.

Once samples are collected they are generally returned to the laboratory as quickly as possible. Sometimes, samples need to be kept anaerobic (without oxygen) or at environmentally relevant temperatures. Other times, we stop all growth and metabolism by flash freezing them in liquid nitrogen at -80°C, and transporting them frozen back to the lab for further analysis.

Evolution of Life on Earth and the Search for Extraterrestrial Life

Yellowstone's geothermal environment is a keyhole

▲ PHOTO 9.7 Thermal features can take on an otherworldly appearance. It is possible that discoveries made in Yellowstone will lead to understanding lifeforms on other planets. (Cathy Whitlock, MSU)

through which we can potentially view the evolution of life from its very beginnings on Earth. The hot springs are possibly more similar to earth just after the inception of life than any other environment. While the organisms that live there have changed and evolved through the eons, they have a record in their genomes of the changes that have occurred. The ways in which they interact with each other and with their environment may give us insights into how early cells, in the absence of oxygen, with an atmosphere of noxious chemicals, may have made their living and began to shape the planet into what we know today.

Early microbes are generally thought to be responsible for generating the oxygen-rich atmosphere that has made earth hospitable to all other subsequent life forms. But research in Yellowstone is showing that they may also have played a significant role in forming the geology of the planet as we see it today as well. These sorts of studies will enable us to look at similar formations if we find them on meteors or on other planets and more easily recognize signs of life.

Understanding these processes may also give us clues as to what to look for as we venture farther out into our solar system and beyond in the search for evidence of life elsewhere. It seems highly unlikely that our first contact with extraterrestrial life will be the green men in space ships or even intelligently formed messages via radiowaves from deep space. More likely, we will find evidence of biological processes that may have ceased millennia ago. Evidence of biologically driven geological depositional processes in the water channels on Mars might tell us of the previous existence of life elsewhere in our solar system. Fossils of mineralized filamentous organisms might be recognized because of our knowledge of similar organisms in Yellowstone and the processes that lead to their mineralization and fossilization.

Back on Earth, it is not unreasonable to expect big things from thermophiles. As the planet warms, our ability to practice agriculture will inevitably change. Previously fertile regions will become too hot and dry to support crops; farming will move into new regions. Meanwhile scientists look for ways to grow crops in warmer, drier environments. Hot springs panic grass, common in geyser basins throughout Yellowstone, can withstand temperatures up to 140 degrees Fahrenheit. However this heat tolerance is the product of a three way mutual relationship between the plant, a stringy microscopic fungus on the plant's roots and a virus that infects the fungus. Thermotolerance imparted on the plant via the virally infected fungus could help create heat tolerant crops for a warming environment. It's not inconceivable that thermophilic microbes and the viruses that infect them may play a significant role in future food and energy supplies well into the next century.

Yellowstone's thermophile ecosystems, it turns out, have a great deal in common with the more familiar macro ecosystem. Each displays an amazing array of biodiversity and interconnectedness. Both have adapted to their environments in complex ways by evolving coping mechanisms to climate and changing conditions. Both hold promise for producing an even wider assortment of ecosystem services beneficial to humans. Already, thermophiles have helped solved murder cases; we anticipate they will eventually improve our lives in other ways as well.

Chapter 10

The Science of Storytelling: Policy Marketing and Wicked Problems in the Greater Yellowstone Ecosystem

Chapter 10

Some issues in Yellowstone seem to never go away. Since the great fires of 1988, various groups have argued about park wildfire management policy. Snowmobile use in the park has been on the agenda since the publication of the winter use plan in 2000. Several lawsuits and millions of dollars later, it is still not settled.

Wolf recovery politics, bison management, and recreation impacts are examples of what Elizabeth A. Shanahan and Mark K. McBeth call "wicked problems." These are issues whose policy setting is characterized as circular, hostile, and unstable. Resolution, if it happens, is short-term and tenuous. Wicked problems are the sorts of politically charged issues that can be manipulated by images and "frames" around which rhetoric and emotion are constructed. Dramatic photos or rhetoric claiming an environmental crisis or threat to private property rights supercharges the debate.

Social scientists describe human behaviors and the study of social life between groups or individuals. Shanahan and McBeth work in an area of policy change, which accounts for the inter-relationships between economic activity, politics, popular culture, and environmental policy. Of particular interest to them is how various groups think about an issue and how they relate to other groups that hold opposing positions. Both sides use language and symbols. Often, the media is the vehicle for these interactions.

The work is important because it helps us understand decisions and social relationships beyond simple partisan politics. A wonderfully simple example is their work on the meaning of YNP. Depending on whether you see the park as a resource to be protected (conserved) or a resource for human use (exploitation), the language used to reconcile the two views becomes important. In the end, both sides may be able to agree on solutions if they can share language designed to minimize, rather than elevate, political tension.

J. Johnson

Wolf watchers at Slough Creek (NPS, Yellowstone National Park)

Chapter 10

The Science of Storytelling: Policy Marketing and Wicked Problems in the Greater Yellowstone Ecosystem

Elizabeth A. Shanahan and Mark K. McBeth

Elizabeth A. Shanahan, Department of Political Science, Wilson Hall, Montana State University, Bozeman, MT 59717; Email: shanahan@montana.edu

Mark K. McBeth, Department of Political Science, Idaho State University, Campus Box 8073, Pocatello, ID 83209; Email: mcbemark@isu.edu

Policy issues in the GYE are a turbulent confluence of divergent human values, contested science, overlapping administrative jurisdictions, and conflicting problem definitions. As such, the study of the complex and dynamic political nature of policy issues in the region is both a highly frustrating and thought-provoking endeavor.

Some of the seemingly intractable policy issues include how the public land agencies propose to manage area wildlife, commodity production, and recreation. Proposed solutions are often confronted with objections by an opposing coalition of stakeholders, which, in turn, lead to protracted lawsuits, lengthy administrative rule making, and dramatic appeals to public opinion by all sides. Many scholars describe policy making in the GYE as "wicked" in nature- circular, hostile, and unstable. In other words, the wicked policy environment of the GYE is one where the process repeats itself several times on one issue, victory is often short-term and always tenuous, and solutions rarely solve the problem; sometimes a solution makes things worse.

The broad aim of science is to build knowledge in the social, natural, and physical worlds. In a politically charged policy environment like the GYE, the challenge for the political scientist is to try to make sense of what is happening and why. Whereas a wildlife biologist might ask, "What is the carrying capacity for bison in Yellowstone National Park?", a political scientist asks, "How do we make decision about managing bison, how do we value them, and how do we understand our relationship to them?"

The policy world is full of people and groups (stakeholders) doing the work of government or trying to influence governmental decisions. The study of political science and policy, famously defined by Harold Lasswell as "who gets what, when and how," is centered on power. While traditional indicators of power in political science include measures such as money and access to decision makers, these aspects do not wholly explain why many of the GYE policy issues are so highly contested. Our research is aimed at achieving an empirical understanding of the wicked nature of these issues. Our approach is different. We focus on understanding political narratives - documents that have been publically disseminated by various stakeholders. We identify the conflicting values or policy beliefs that propel contested battles and analyze the political narrative tactics that are used to influence policy outcomes.

Policy making in the GYE does not occur in a vacuum or somehow separate from larger social and political realities. Instead, policy conflict in the GYE is reflective of policy problems and processes facing American democracy. Thus, given the political and social context of the wicked policy environs of the GYE, we can get a glimpse into the health of our democracy and what our responsibilities are as citizens to contribute to our democratic wellbeing.

To best illuminate how we conduct our social science research, we begin with detailing the historical and cultural basis for divergent value differences in the GYE. We then specify how we conduct our research to understand the wicked nature of the policy environment. We conclude with reflections on how both our research approach and our results help us know Yellowstone. Finally, we give you an opportunity to test drive our methodology.

The Greater Yellowstone Area in Context

One of the reasons there is such a wicked policy environment in the GYE is due to divergent perspectives as to the meaning of Yellowstone itself. When the Yellowstone National Park Act (1872) was passed and created the world's first national park, it set the stage for a contested meaning of what Yellowstone National Park (YNP) is or would become. The statute articulates the dual mission of YNP is to be both a "pleasuring ground

▲ PHOTO 10.1 Roosevelt Arch at the North Entrance commemorates the creation of Yellowstone National Park in 1872. The original mission of the National Park Service is "...to promote and regulate the use of the...national parks...to conserve the scenery and the natural and historic objects and the wildlife therein and to provide for the enjoyment of the same in such manner and by such means as will leave them unimpaired for the enjoyment of future generations." Even during the fires in 1988 the Park stayed open to visitors so they could experience wildfire firsthand. (NPS, Yellowstone National Park)

for the benefit and enjoyment of the people" and a place reserved "for the preservation, from injury or spoliation, of all timber, mineral deposits, natural curiosities, or wonders within said park, and their retention in their natural condition." Following the creation of YNP, various meanings of the Park have evolved. In turn, these meanings form the basis of policy debates of the Park's management practices.

The birth of YNP was built on a "creation myth," which framed the meaning of the Park as a natural wonder to protect. The classic story is that the Washburn expedition camped at Madison Junction in the fall of 1870 and decided that because of Yellowstone's immense size and beauty, the region should be set aside as a national park. However, Yellowstone historians Paul Schullery and Chris Magoc find the diaries made no mention of such a conversation even though members of the Washburn party wrote of many far less momentous discussions. The covert propagator of the creation myth story is believed to be Nathaniel Langford, who in 1905 was a supporter and advocate of the Northern Pacific Railroad. In fact, the passage of the Yellowstone National Park Organic Act economically advantaged the railroad when Northern Pacific officials successfully lobbied for creation of the park. The railroads wanted a western tourist attraction and recreational opportunities as an impetus to extend their rail lines to the American West for their own economic gain. These two meanings of YNP- a resource to be protected (intrinsic value) and a resource for human use (economic and recreation value) - dominate the current political environs of the GYE.

Meaning attributed to something, whether it is YNP or the American Flag, is based on normative values. Normative values are personal beliefs of how things *ought* to be or how things *should* be. For example, in the environmental politics literature, how people view the relationship between humans and nature is a normative value that drives how people ascribe meaning to nature. An anthropocentric normative view (sometimes referred to as cornucopian or conservationist view) of this relationship understands nature's purpose as fulfilling human need for survival and progress; a biocentric normative view (sometimes referred to as preservationist or intrinsic value view) tends to see nature as harboring an intrinsic value, where human needs are no greater than those of nature. This normative debate is often portrayed through the comparison of Gifford Pinchot's conservationist perspective with that of John Muir's preservationist perspective. Since the early western expansion days, these opposing meanings for our public lands have been hotly debated.

In the GYE today, the cornucopian perspectives and environmental protection values are exacerbated by a rapidly changing American West. Deep cultural divisions center on competing symbolic meanings of the American West that reflect both the romantic preservationist perspective and rugged individualism economic perspective. This regional cultural division has intensified given recent rapid changes in the GYE's economic and social landscape. The local economies of the American West have experienced a decline in its extractive-resource base (e.g., mining, logging) and an increase in three economic sectors: tourism, construction, and service-based industries. The changes in the economic base and increases in population have translated into cultural differences that are typically cast in the dichotomy of the Old West (a rural, agricultural, and natural resource based economy) versus the New West (an urban, environmental amenity - service based economy fueled by tourism). This Old West - New West dichotomy, while overly simple, is useful for understanding how policy stakeholders in the area have coalesced around similar meanings of YNP and its surrounding area.

Actors in the Policy Process

Stakeholders in the policy process are those individuals or groups who have an interest in a particular policy outcome. There are a plethora of stakeholders in the region, and they vary according to the policy issue at hand. Each of the individuals within these groupings harbors a normative perspective regarding the meaning of Yellowstone and each holds different values and divergent policy beliefs. In turn, the policy arena is increasingly polarized.

INTEREST GROUPS

Interest groups typically represent a well-defined constituency of like-minded members willing to incur some cost of being included in the group. Interest groups in the area tend to cluster into one of three categories: environmental interest groups (e.g., the Greater Yellowstone Coalition, Buffalo Field Campaign), motorized recreation groups (e.g., Blue Ribbon Coalition), and extractive industry groups (e.g., Montana Stockgrowers Association; Montana Logging Association). The environmental groups' idea of resource protection is one that places YNP in the context of the larger GYE ecosystem as one of the last vestiges of untouched wilderness in the contiguous United States. Some environmental groups believe that Yellowstone bison should be free to migrate beyond Park boundaries in the winter months, despite the concern of ranchers that these bison may transmit a disease (brucellosis) that would gravely affect the cattle industry. Motorized recreation groups see the GYE as an area reserved for motorized recreational access throughout all seasons; these groups advocate for snowmobile access to the Park over objections that the machines pollute and stress wildlife during the hard winter months. The extractive industry groups view the economic viability of their livelihood as having precedence over protection of the resource goods in the area. They believe the sustainability of logging is more important than protecting grizzly habitat through the designation of roadless areas in national forests.

CITIZENS

Citizens are a broad category of stakeholders that include visitors to the area and gateway community residents. First-time visitors to YNP often have a conception of the Yellowstone as a zoo. As Lee Whittlesley and others have pointed out, tourists often view Yellowstone wildlife not as real creatures but rather as symbols or relics of times gone by. The boundaries surrounding the Park are understood as real and meaningful, thus creating an inside-outside dynamic that excludes any notion of ecosystem management and biodiversity. To those who live in the gateway communities surrounding the Park, Yellowstone is often seen as a commodity that allows them to make a living. To achieve this end, they view the Park as a place to protect, but also as a place for mostly unregulated access so that the visitors will sustain the local economies by eating and seeking lodging in the gateway communities.

ELECTED OFFICIALS

Different governing coalitions have engendered varying meanings of the GYE. For example, presidential administrations have weighed in on GYE policies that overturn previous administrative decisions and ricochet issues such as snowmobile access across bureaucratic and judicial jurisdictions. The Clinton era engendered a protectionist meaning for YNP, whereas the Bush administration saw YNP as a resource for humans (recreational, economic). However, it is both correct and overly simplistic to argue that partisanship plays a role in GYE politics. For example, Republicans tend to favor extractive commodity use and reject regulation; thus, it is not surprising that the Wyoming and Idaho Republican congressional delegation have been united in opposition to wolves and snowmobile regulation. However, while Montana Republican governors Marc Racicot and Judy Martz strongly supported the lethal management of bison outside of Yellowstone, the issue has hardly been resolved under the administration of Democrat Brian Schweitzer. Furthermore, the late U.S. Senator Craig Thomas (R-Wyoming) was lauded by the environmental interest group the Greater Yellowstone Coalition for his opposition to natural gas development on public lands. Montana Democratic U.S. Senators Max Baucus and John Tester rarely take the lead on GYE environmental policies.

BUREAUCRATIC AGENCIES

Public agencies constitute another group of stakeholders who harbor meaning for what Yellowstone is. Many of the career public servants working in these agencies have scientific expertise in understanding YNP and the GYE. For example, the National Park Service (NPS) has biologists on staff and the Montana Department of Livestock (MDOL) has a state veterinarian. Yet, these agencies cannot help but to feel the pressures from their institutional cultures and political overlords. They are thus influenced by the policy desires of varying administrations (presidential and gubernatorial) as well as local and regional concerns, needs, and powers. At times, the meaning of Yellowstone is divided between their role as scientists and experts and their role in the fragmented political world of policy making. For example, the NPS strongly supported the elimination of snowmobiles inside YNP because of concerns over pollution and wildlife stress, but local politics and presidential orders have required them to accommodate snowmobile access. Similarly, the MDOL wants the Yellowstone bison to remain within YNP boundaries over the concern for spreading brucellosis to cattle; recent cases of the disease in cattle herds have expanded the attention to include wild elk herds. Thus, there is no coherent ecosystem policy in part because other federal and state agencies, compared to the NPS, have different constituencies, legislative mandates and institutional incentives. Because of the conflicting demands and lack of coordination among agencies, the NPS finds it difficult to execute a scientifically based plan for the management of Yellowstone.

In sum, from the National Park Service enabling legislation in 1872, to the recent economic and cultural changes, the meaning of the Park and its surrounding area cannot be reduced to one 'correct' meaning for stakeholders in policy decisions. The central battle over the meaning of the GYE is one of economic use versus

▲ PHOTO 10.2 Winter recreation in Yellowstone is a "wicked problem" that has been debated since the Park was opened to snowmobile traffic in 1974. Park Service concerns about the effects of snowmobiling prompted limits on the number of machines allowed to enter the Park. Limits have been implemented, overturned, and reinstated in recent years. The use of the political system for this wicked problem runs the gamut of executive orders, judicial rulings, and promulgated rules. (NPS, Yellowstone National Park)

environmental preservation. The wicked or intractable nature of policy making in the region means that these policies will not be resolved by factual argument. In such controversies, long-standing policy solutions rarely occur because policy battles tend to revolve around what Advocacy Coalition scholars refer to as *core policy beliefs* or normative values that groups or individuals hold as 'true' rather than objective scientific fact. It is upon these core policy beliefs that the various meanings of YNP and the GYE are built and wicked policy ensues.

Wicked Policy Research in the Greater Yellowstone Area

Social scientists, like any scientists, try to explain why certain phenomena happen the way they do. In our case, we seek to understand the wicked nature of policy issues: *why do problems that are seemingly solvable using science, common sense, and practicality escalate to high levels of hostilities that lead to lawsuits, stalemate, and solutions that serve the interests of neither warring party?* Our studies specifically focus on how stakeholders' policy narratives contribute to policy wickedness in the GYE. Below, we present the findings from three studies of policy narratives.

POLICY MARKETING: IS PUBLIC OPINION FOR SALE?

Much of our work is an empirical analysis of how policy narratives intensify political conflict and gridlock. But, in order to understand these results, we must first describe a theory for why policy narratives are so powerful. Our theoretical framework places the GYE into the context of larger societal trends - that of the entrenchment of a marketing culture and the rise of consumerism. Our argument is that the contemporary economy no longer focuses on production but rather on creating demand for goods and services. This shift has led to a marketing culture, a culture that has permeated political life leading to the "permanent campaign" and to marketing slogans being used to sell everything from environmental policy (e.g., the Bush administration's "clear skies" initiative) to education (e.g., No Child Left Behind) to war (e.g., "shock and awe," and "the surge"). Concurrent with this trend has been the development of a "consumer" culture or orientation among citizens. For many, a person's identity comes not from their job but from their consumption. Citizens become passive recipients of marketing symbols and emotional sound bites, not only in the economic market, but also in the political system. We argue that the consumer's political knowledge and interests are marketed to them and policy marketers, not citizens, define public policy problems. The ensuing policy solutions are related more to ephemeral lifestyle choices than they are to rational debate or economic and political interests. This theory of policy marketing occurs in the GYE, as interest groups, elected officials, and the media all engage in activities that seek to shape public opinion and contributing to policy intractability. Like any ad campaign, these narratives sell new ideas, new crises, and new solutions.

Our next step provides empirical work to support our claim that policy narratives are purposefully constructed by groups to influence public opinion and policy outcomes. Thus, we pursue two broad research questions: 1) what do these policy narratives look like? (in other words what are the empirical elements of a policy narrative?) and 2) is it possible to quantitatively study policy narratives using existing policy theory? Research questions arise from theory and guide the research. Developing research questions is key to all scientific work. The balance of this chapter presents the process by which we investigate these two research questions - something we call "the Science of Storytelling."

THE SCIENCE OF STORYTELLING PART I: CORE POLICY BELIEFS

One of the mainstream theories that inform our work is Paul Sabatier and Hank Jenkins-Smith's Advocacy Coalition Framework, which asserts that groups hold core policy beliefs or normative values that serve as the glue for advocacy coalitions. When disputes between opposing coalitions center on these beliefs, they lead to conflict. The result is a lack of policy learning between groups as accumulated scientific evidence is ignored in favor of value-based conflict. The empirical measurement of these beliefs has been problematic in the literature, and we wanted to test whether narratives found in stakeholders' documents were a reliable source of policy beliefs. Thus, our first investigation was testing whether there was a "science of storytelling" or policy beliefs predictably and consistently embedded in stakeholders' policy narratives.

There are three core policy beliefs critical in the GYE: federalism, the relationship between humans and nature, and type of science used.

FEDERALISM. Federalism is the normative belief of what level of government should make the policy decision—the federal government (national federalism) or local governmental entities (compact federalism). Differing views of American federalism play an important role in Yellowstone politics. Old West groups contend that GYE issues are ones that affect local citizens; therefore, locally elected officials and local groups should have the power to decide policies. New West groups are more likely to state that Greater Yellowstone is an area of national concern and national groups, citizens, and elected officials outside the region should be central to policy making.

RELATIONSHIP BETWEEN HUMANS AND NATURE. This core policy belief is best understood in terms of divergent management orientations. The New West philosophy is reflective of biocentricism, where human interference in natural processes is kept to a minimum and natural resource extraction takes a back seat to biodiversity. Conversely, the Old West philosophy is based on active natural resource management and a belief that humans rule nature for our purposes.

SCIENCE. The differing role of science is a key issue in Greater Yellowstone debates. The science preferred by New West groups is characterized by natural management, habitat and ecosystem protection, and biodiversity. In addition, conservation biology that favors biodiversity and protection of endangered species is a popular scientific position of New West advocates. Old West groups, on the other hand, argue that technology can correct environmental problems. They value science that is human centered; they view nature as a commodity that is to be directed and managed through technological innovation.

▲ PHOTO 10.3 Content analysis begins with assembling public documents, seen here as newspaper articles from local Greater Yellowstone Area media outlets, and meticulously reading the narrative, coding the contents, and recording the results into a codebook. Here are two researchers reconciling their individual coding results of press coverage of wolf, bison, and snowmobiling wicked policy issues to ensure inter-coder reliability. (Liz Shanahan)

These policy beliefs are all around us, in press releases, newsletters, and media spots. The issue is - are they embedded in New West and Old West policy narratives? Because these policy narratives are everywhere, with the potential to be very influential on our opinions, they are important sources of stakeholder belief systems.

We investigate the science of storytelling by identifying all public documents generated by two prominent interest groups in the region — the Greater Yellowstone Coalition (GYC) and the Blue Ribbon Coalition (BRC). We chose these groups because they represent vastly different constituencies, environmentalists and motorized recreationalists, respectively and, seemingly different values with respect to nature. The documents - press releases, newsletters, and editorials - represent all known public documents released by the both interest groups over six years. The requirement was that the document had to address one of three policy controversies: bison management, the Clinton Administration roadless initiative, and snowmobile use in Yellowstone National Park.

Each document was content analyzed for core policy beliefs; documents were read and coded for what kind of policy belief was found in the narrative. Coding is a methodological tool for systematically identifying patterns in the text. In this case, federalism, human/nature relationships, and type of science used in the narratives is language that is crafted to frame the issues in ways that produce desired political outcomes. As Deborah Stone argues, problems are defined strategically and are constructed by influential individuals or organizations as part of a political game. Problem definitions are embedded in narratives through the use of literary devices such as characters, plots, colorful language, and metaphors. As Stone points out, the ultimate goal is to define a political problem so that your side's stance appears to be in the public interest.

The first two policy beliefs were identified through strategic use of character casting. Those entities cast as heroes revealed what perspective of federalism a narrative held. For example, if the President was cast as the hero in addressing the issue then it showed national federalism; if it was local officials, then it was compact federalism. Those cast as the victim revealed the policy belief of human-nature relationship; if nature (i.e. bison) is the victim, then it shows a biocentric policy belief; if snowmobilers are the victim, then the narrative shows an anthropocentric policy belief. The scientific evidence cited in narratives was coded as one that offered a technical fix or one that had a conservation perspective. We developed a codebook to document what narrative elements were in each document. We did not use computers to scan the documents; rather, we trained graduate students to read and code each document. After they each read the documents individually, they then compared what they coded to ensure that the coding was accurate (called inter-coder reliability). We entered the data into a statistical program and were able to determine if there were statistical differences in policy beliefs between the two groups.

The research team found that the environmental group harbored the following core policy beliefs:

- national federalism policy belief;
- biocentric in their orientation of humans and nature;
- use of a mixture of conservation biology and technologically based science to support their desired policy outcome.

In statistical contrast stood the motorized recreation interest group. Their narratives held the following core policy beliefs:

- local or compact federalism policy belief;
- anthropocentric orientation of the human and nature relationship;
- exclusive use of a technological or anthropocentric science as evidence.

The importance of this work is that indeed policy narratives have embedded policy beliefs and that these beliefs reveal a statistically significant difference between groups. Thus, when the opposing coalitions talk about policy issues in the GYE, they do so through different lenses, and the battles between coalitions and their citizen audience become as much about what policy beliefs should be dominant as they are about the problem itself.

THE SCIENCE OF STORYTELLING PART II: NARRATIVE POLITICAL TACTICS

The next step in our research centered on whether policy narratives had embedded narrative strategies to try to influence policy outcome in the GYE. In politics, there are winners and losers. As E.E. Schattschneider describes, winners try to contain the policy issue by controlling the number of groups affected by the policy in order to maintain the status quo. Losers try to expand the arena of conflict to more groups to gain policy support. In other words, a coalition that perceives itself as losing will try and bring in other players to the coalition to build salience, and a coalition that perceives itself as winning will try to restrict participation. The net effect is to perpetuate the wicked nature of policy conflict.

In the policy studies literature, three political narrative strategies and tactics are identified: concentration of benefits and costs of proposed policy outcomes, condensation symbols, and policy surrogates. These are the theoretical constructs for which we developed a methodology.

The Basics of Narrative Content Analysis

The goals of a policy narrative include:

(1) constructing a group or coalition identity; (2) defining the group telling the story as heroic; (3) constructing victims that resonate with audience members; (4) constructing the narrative of the group as the public interest; (5) finding despicable villains that are opposed to the public interest; and (6) either trying to expand or contain policy conflict, depending on whether the group sees themselves as winning or losing.

Characters (heroes, villains, and victims) are critical elements of a political narrative. Below are definitions and examples from two competing interest groups in the GYE surrounding the issue of snowmobile access to YNP- a motorized recreation group (MRG) and an environmental group (EG). We have written fictionalized quotes to give you examples of how the two opposing sides might portray their narratives.

▶ *HEROES:*
the group telling the story and allies of the group are heroes; they are the potential fixer of a problem.

MRG: "Our local communities are working hard to stop the Clinton madness."

EG: "We have many allies in Washington, DC working behind the scenes to help us stop the Bush administration from repealing the snowmobile ban."

▶ *VILLAINS:*
the group or person that is causing the problem (normally constructed as opposing the public interest); villains benefit from other groups' suffering.

MRG: "Environmental groups are working with their Hollywood friends to reverse the hard-won grassroots victories of our group."

EG:" Snowmobilers are going off-trail and harming rivers, plants, and wildlife."

▶ *VICTIMS:*
person(s) or things harmed by the villain and pay the costs of another group's actions.

MRG: "Local communities are harmed by the Clinton rule! I can see businesses closing in West Yellowstone."

EG: "Rangers are getting sick from snowmobile fumes and visitors complain that they cannot hear the wild sounds of Yellowstone over the noise snow machines."

BENEFITS AND COSTS. Winning groups seek to contain the issue by concentrating costs of the proposed policy solution (i.e., few carry the financial or political burden) and diffusing its benefits (i.e., many winners), whereas losing groups diffuse costs (i.e., many carry the burden) and concentrate benefits (i.e., few winners).

CONDENSATION SYMBOLS. Groups like to take complex issues and simplify them by attaching easily understood symbols to the policy issue. The goal is to make complex issues easy to remember. We empirically tested whether losing groups use condensation symbols more than winning groups, given that this strategy could expand the issue to larger constituencies. Examples of condensation symbols are environmentalists referring to YNP as a "motorized race course" or to elected officials as "corporate politicians."

THE POLICY SURROGATE. As Martin Nie and others point out, in wicked policy environments, groups complicate issues by linking them to larger issues. For environmental policy in the American West, this means that issues like bison management and snowmobile access are wrapped in larger controversies such as concerns about federalism and the fear of outsiders. We investigated whether losing policy narratives strategize by using policy surrogates to entangle policy issues in larger, emotionally charged debates. They would do so in an effort to gain a competitive advantage by expanding the scope of the policy issue.

Having identified tactics used in other political debates, we developed a second codebook to systematically identify policy strategies in political narratives. We collected eight years of the same prominent interest groups' policy narratives and content analyzed them for the three political narrative strategies. In contrast to our previous study that hypothesized the differences in core policy beliefs to be between interest groups, we hypothesized in this study that the differences in narrative political strategies would be between those who believed they were winning or losing in the policy battle. So, despite differences in core policy beliefs, we hypothesized that interest groups would use the same narrative political strategies to achieve policy success. Indeed, this is the case.

We found that part of the wicked policy environs is characterized by the two groups portraying themselves in their public documents as losing 68% of the time. Given this preponderance of losing narratives, the groups were far more likely to be expanding the arena of conflict through their narrative tactics than they were to be trying to contain the arena of conflict. Across all group policy narratives, those that posited themselves as losing used the following policy narrative strategies:

- Both interest groups *concentrate benefits* in order to construct a situation where it appeared that only a small group benefits from the status quo. For example, the Blue Ribbon Coalition argues that only environmental groups benefit from the banning of snowmobiles. Similarly, the Greater Yellowstone Coalition promotes the idea that President Bush ignored national interests in favor of the snowmobile industry.

- Both groups *diffused costs* when losing in order to demonstrate that many are harmed by the status quo. The Blue Ribbon Coalition tends to focus on how individual snowmobile users, communities, and the economy loses when snowmobile use is regulated. Likewise, the Greater Yellowstone Coalition argues that snowmobile use harmed a variety of factors including human health, wildlife, and visitors.

- When losing in the policy debates, groups use *condensation symbols* to try to expand the policy issue to a larger audience by evoking an emotional investment in the policy issue. Here we found that the Blue Ribbon Coalition would often go after environmental groups directly labeling them as elites. Conversely, the Greater Yellowstone Coalition often defines the issue in terms of Yellowstone as a speedway.

- The *policy surrogate* is a narrative tactic that tries to expand a policy issue by connecting it to a larger and often highly contentious issue. Here, we found that the Blue Ribbon Coalition often focuses on federalism (in essence, arguing for local control over federal dominance) and for democracy and grassroots policy. The Greater Yellowstone Coalition is more likely to focus on corruption and special interests and contrast industry with the importance of Yellowstone as the United States first national park.

Taken together, we learned that in the wicked policy environment stakeholders tend to portray themselves as losing and use consistent political narrative strategies in an effort to expand the arena of conflict. The persistence use of such narrative strategies further escalates policy conflict in GYE.

▲ PHOTO 10.4 Interest groups use the media as an inexpensive way to get their policy message out to a broader public. The question is - do the media serve as a conduit for various opinions on wicked problems or, does it act as a conductor of policy opinion by framing issues with a particular policy outcome in mind? (Buffalo Field Campaign)

IS THE MEDIA A STORYTELLER?

Thus far, through our coding schemes we found that environmental and motorized recreation interest groups do generate narratives that harbor internally consistent policy beliefs that are statistically different from the other group. We have also found that these groups utilize similar political narrative strategies when they portray themselves as losing, despite having divergent policy beliefs. But what of other stakeholders in the GYE? What of the media?

The media is supposed to be objective, and yet there is often accusation of bias - both nationally and regionally. We wondered to what extent the media are a conduit for public opinion or a contributor or active participant in policy debates. To answer this question, we decided to test whether newspaper articles are policy narratives, with embedded policy beliefs and media framing strategies.

▲ PHOTO 10.5 Under the Interagency Bison Management Plan, the Montana Department of Livestock has the authority to haze bison back into the Park boundaries using horses and helicopters. Some see this solution as better than slaughtering bison, others claim that the hazing is aggressive and provokes unnecessary anxiety and injuries to bison. (U.S. Fish and Wildlife Service)

Due the national attention given to GYE issues, we designed a study to compare eleven years of media coverage of two GYE issues (wolf reintroduction and snowmobile access) in both national (*USA Today* and *New York Times*) and local (*West Yellowstone News* and the *Cody Enterprise*) news papers. We were able to rather easily obtain the national articles through a search using LexisNexis. However, since local newspapers do not have resources to create digital archives of past issues, we conducted archival research. This means that we went to local libraries and spent dizzying hours looking through microfiche and yellowed originals for coverage of the two policy issues. Archival research is a time-intensive process, and one reason most research focuses on national media coverage.

In tandem with collecting all relevant articles, we designed a new code book that would capture both policy beliefs and narrative framing strategies. To code for policy beliefs in media articles we utilized "source cues". These are the people and groups from whom journalists gather information; these source cues were identified and coded. We found that national newspapers had statistically higher rates of national based source cues; we concluded that, like the Greater Yellowstone Coalition, they possessed a national theory of federalism. Conversely, the local papers primarily depended on local source cues and hence, like the Blue Ribbon Coalition, exhibited a compact theory of federalism. Thus, we found that the local and national media, through their source cues, are active contributors to the GYE policy debate by harboring different core policy beliefs of federalism.

However, on the remaining two policy beliefs (human-nature relationship and type of science used), there were no statistical differences between national and local coverage. Similar to our work with interest group narratives, we coded whether the victim was anthropocentric (e.g., saving Yellowstone for humans) or biocentric (e.g., saving nature). Unlike their interest group counterparts, *both* the national and local papers portrayed the victims of the policy issues in anthropocentric terms. Also surprising was that the media cited science very infrequently, leading to an inability to interpret he quality of the science they used.

In terms of narrative framing strategies, the nationally and locally based source cues were coded for New West or Old West policy orientation. In other words, source cues tended to espouse a policy leaning, pro- or anti-snowmobile access and pro- or anti-wolf reintroduction. In turn, these policy leanings were grouped into two categories: Old West (pro-snowmobiling and anti-wolf) and New West (anti-snowmobiling and pro-wolf). The national papers used New West (pro-wolf, anti-snowmobile) source cues and the local papers used Old West (anti-wolf, pro-snowmobile) source cues. Likewise, national papers use New West descriptors of the wolf such as "noble", "culturally precious", and a "cuddly favorite"; in contrast, local articles highlight Old West descriptors of the wolf such as "strong predator", "the AIDS virus", and "abusive and arrogant". Such descriptors found in media accounts reveals a narrative strategy meant to influence public opinion.

Again, the major question asked in this phase of our investigation is whether the media is simply a conduit of information (reporting multiple policy preferences in newspaper accounts) or whether they are a contributor in the policy debate (constructing policy stories that harbor consistent policy beliefs and narrative framing strategies). We found a nuanced policy landscape. Rather than the view of the media as either an advocate or a conduit we found a mixed role for both national and regional newspapers. They are a contributor in particular instances and a conduit in others. We conclude that the media is a factor in the wicked policy environs of the GYE by offering often incomplete policy narratives with a strong dependence on framing but poor use of science in their coverage.

As political scientists, we study power, democracy, policy change, and policy gridlock. While more traditional political scientists would approach knowing Yellowstone through a study of institutions, legal matters, or public opinion, we choose to study the core of politics the fragmented, messy, and ever-evolving political world found in the language or narratives used by political actors. So what have we learned?

At each step of the way, our research is grounded in current theory to learn more about the wicked policy process. To build knowledge through research requires knowing and testing the discipline's theory. In this case, we were able to find evidence for theoretical assertions that had never been tested in real world application. Second, our research is an example of how the process is iterative in nature. In other words, building knowledge consists of small steps that build on one another to continue to fill in the picture of the phenomenon under study - in this case wicked policy problems. Research rarely follows a linear path and ours certainly did not. We jumped from advocacy groups, to media studies, and now intend to study elected officials and bureaucratic agency workers as policy actors. Third, our research is necessarily systematic and transparent. These are critical aspects of any research, so that others can repeat the study and either confirm or deny our findings. Additionally, while most data comes in the form of numbers, our data comes in the form of words. Given the subjective nature of interpreting words, it is all the more crucial to be systematic and consistent. Finally, all research must answer the "so what" question. So what if there are different coalitions using narratives to make the GYE policy arena a wicked one? So what if the media participates in these policy battles? So what if the GYE has a changing culture? This is where it is critical that we place our results in the larger context of democracy. To do so gives import to our work beyond the boundaries of the GYE policy environs.

Our studies have confirmed the wickedness of the GYE policy environs. All stakeholders are losing or at least think they are losing power and thus view politics as a zero-sum game. Policy marketing is equivalent to soft drink commercials where Pepsi never acknowledges or discusses the positive attributes of Coca-Cola. Instead, policy making, just like niche marketing, is a competitive game with winners and losers. Both sides have power but neither side can dominate the other. Often, the goal of each coalition is not to win but rather to keep the other side from winning.

Interest groups ground their policy narratives in fundamentally different policy beliefs and use narratives to expand the arena of conflict. They use provocative metaphors and characterize their opponents in the most unflattering terms possible. Specific issues like bison management or snowmobile regulation are quickly wrapped up in larger cultural issues, such as "rugged individualism" or federalism. The media, likewise, gets involved in this game. The national media sees Yellowstone as a national issue and the local media sees it as a local issue. Both sets of media use descriptors and metaphors that reflect their respective policy interests.

▼ PHOTO 10.6 Managing bison in Yellowstone has been controversial since the Park's early days. In the 1920s, the wild bison population was augmented with domesticated animals. When the bison herd was determined to be too large, the herd was reduced through the late 1960s. Natural ecological processes are now used to determine bison numbers and distribution but the brucellosis issue has increased pressure to once again control bison numbers artificially. (Buffalo Field Campaign)

▲ PHOTO 10.7 Wilderness, multiple use, and recreation are all legitimate uses of public land in the Greater Yellowstone Ecosystem. Public land managers face a complex balancing act between managing for human enjoyment and preservation of natural ecological functions. Collecting meaningful public input on the future of public lands can be a significant challenge for land managers. (Jerry Johnson)

The issue with value-based conflict is that cultural values often anchor GYE groups to positions that appear to be, in fact, *opposed* to their fundamental interests. For example, in the bison controversy, it can be argued that the economic and power interests of ranchers were not served by the actions of Montana elected officials. They argue for local control over the killing and testing of Yellowstone bison in lieu of federal intervention. Local control led to an unintended consequence as the bison management program mobilized a wide-scale national reaction in favor of bison and against the rancher's own perceived interests. Similarly, environmental groups who vilify local ranching communities seem to be alienating potential allies for environmental preservation. A cultural value of New Westerners seems to be the need for amenities. Yet, population growth and the consequential subdivisions have resulted in serious water and land use concerns in the region.

Conclusion

If Yellowstone is our laboratory, then does this wicked policy environment exist outside of the region? We argue yes, and it is spreading, as policy narratives become political weapons in national policy wars. How to dismantle marketed discourses based on values and replace them with authentic discourses based on interests remains a perplexing dilemma.

A starting point for such cooperation would occur in the policy narratives of each group. Such narratives would focus less on policy beliefs and more on the mechanics of policy problems. Such narratives would focus more on interests and less on cultural beliefs and myths. They would not portray the groups writing the narratives as losers or victims; such narratives would seek to contain policy issues to a manageable size of stakeholders. It is possible to imagine a Yellowstone where politics becomes the art of the possible and through cooperation opposing coalitions both win. Maybe, the "other side" is not as evil as policy marketers like to portray them. Maybe, environmental preservation and economic sustainability are possible. Maybe, grizzlies, wolves, bison, elk, and humans can live with some accord. We could change the status quo first through the construction of collaborative narratives then design non zero sum solutions. An important player in these policy battles is the collective citizenry who 'consume' these policy narratives. Citizens in the Greater Yellowstone Ecosystem and elsewhere live in competing social realities. They depend on mutually exclusive sources of knowledge and competing interpretations of that reality. The rise of policy marketing by entities is clearly a subtle manipulation of public opinion. When citizens examine policy conflicts, they, like the policy marketers that provide the information, often approach the conflict from diametrically opposed frames that fail to consider the values of the opposition and the larger context of Greater Yellowstone policy. However, we contend that it is possible and necessary for GYE citizens and others interested in the GYE to think and read critically and not be the subjects of policy marketers and their divisiveness. This is the democratic ideal of an informed citizenry.

Coding a Policy Narrative for Policy Beliefs

Below is a policy narrative from a hypothetical environmental nonprofit. Following the narrative is a code book similar to the ones we use in our work. Code the narrative, fill in the codebook and consider the following questions:
Was it difficult to code? Why?
How do your results compare with ours?
In what ways do you think this political narrative might contribute to the wicked nature of policy on the GYE?

▶ *TITLE:* Yellowstone as NASCAR: How the Bush Administration is Undermining American Democracy
▶ *FROM:* The Green Coalition

The Bush administration and corporate interests are working hard to overturn the Clinton administration's banning of snowmobiles from Yellowstone National Park. The Seattle-based Green Coalition is working with top elected officials and experts to stop this egregious abuse of power. Members of the U.S. Congress are taking notes of the administration's actions. For example, U.S. Representative Moss (D-NY)recently stated on the floor of the House of Representatives that the Bush administration has consistently neglected the will of the American people as well as scientific evidence proving the harm of snowmobile exhaust to park rangers. Furthermore, wildlife scientist Sally Randall argues that wildlife are unduly stressed by motorized recreation as they are chased off the road in the wintertime when their energies are low. The Green Coaltion, in conjunction with the Ecological Foundation of America, is also currently sponsoring a study to determine the impact of winter snowmobiling on plants. According to the Foundation, "snowmobiles harm plants by subjecting them to physical stresses."

The Green Coalition has worked hard throughout the 1990s to eliminate snowmobiling. Our hard work paid off, but now it is all in jeopardy since President Bush is seeking to overturn what is commonly called the Clinton rule. Powerful, elite special interests who will benefit directly from this rule change back the Bush administration's efforts. Some locals, including a former city council member of a local community, however, believe that ending snowmobiling will actually help local businesses. If the administration is successful at overturning this rule, than other rules such as the Clinton Roadless Initiative may well be at stake. Such an overturning will give tremendous power to the new administration in undoing Clinton era environmental polices. This will harm innocents like fisherman and cross country skiers. In fact, American democracy is at risk.

Practice Codebook

Fill in the following codebook for the above policy narrative.

▶ ***CORE POLICY BELIEFS***

1. Who are identified as the allies?

___ of national allies

who:

___ of local allies

who:

2. Who are identified as the victims of the problem presented?

___ nature, wildlife, ecosystem victims

who:

___ human victims

who:

▶ ***NARRATIVE STRATEGIES***

3. Does the narrative take a stance toward the current situation?

___ positive or ___ negative

comment:

4. Are costs of the current problem (i.e., those who will pay for overturning of the Clinton rule) diffused or concentrated?

___ diffused costs or ___ concentrated costs

how:

5. Are the benefits of the current problem (i.e., those who will benefit from the overturning of the Clinton rule) diffused or concentrated?

___ diffused benefits or ___ concentrated benefits

how:

6. Are condensation symbols used in the narrative?

___ yes or ___ no

what were they:

7. Is there a policy surrogate?

___ yes or ___ no

what were they:

Answers to the Practice Codebook

Fill in the following codebook for the above policy narrative.

▶ ***CORE POLICY BELIEFS***

1. Who are identified as the allies?

__5__ of national allies

who: U.S. House Rep Moss, Sally Randall, and Ecological Foundation of America, the Green Coalition, and the Clinton Administration

__1__ of local allies

who: former city council member

2. Who are identified as the victims of the problem presented?

__5__ nature, wildlife, ecosystem victims

who: wildlife, exposed and unexposed vegetation, seedlings, shrubs

__4__ human victims

who: park rangers, the American people, fishermen, cross-country skiers

-- Questions 1 and 2 are constructed to measure two policy beliefs of this group. For each document, we calculate a core policy belief score, that represents a place on the continuum of -1.00 (Old West) to +1.00 (New West).
-- The coding of allies (heroes) is a measure of the stakeholder's policy belief of federalism. The score for this document is put into the following formula: # of national allies - # of local allies / total number of allies
-- Thus, the score for this document would be: (5-1)/6= +0.667 (nationalism; New West)
-- The coding of victims is the measure of the group's policy belief of the relationship between humans and nature. The score for this document is put into the following formula: # of nature and wildlife victims - # of human victims / total number of victims
-- Thus, the score for this document is (5-4)/9 = +0.111 (biocentric; New West)

▶ ***NARRATIVE STRATEGIES***

3. Does the narrative take a stance toward the current situation?

____ positive or __x__ negative

comment: the group is portraying themselves as losing

4. Are costs of the current problem (i.e., those who will pay for overturning of the Clinton rule) diffused or concentrated?

__x__ diffused costs or ____ concentrated costs

how: they are diffused. fishermen, skiers, other rules may be reversed, the American people, vegetation, democracy is at risk

5. Are the benefits of the current problem (i.e., those who will benefit from the overturning of the Clinton rule) diffused or concentrated?

____ diffused benefits or __x__ concentrated benefits

how: the benefits are concentrated, powerful special interests

6. Are condensation symbols used in the narrative?

__x__ yes or ____ no

what were they: Yellowstone as NASCAR; special interests

7. Is there a policy surrogate?

__x__ yes or ____ no

what were they: American democracy is at risk

-- This is a classic "loser's appeal" using a stymied progress story aimed at expanding the arena of conflict. While portraying themselves as losing, a group will tend to: diffuse costs, concentrate benefits, and use condensation symbols and policy surrogates.

▶ Grand Prismatic Spring, Yellowstone National Park (NSP, Yellowstone National Park).

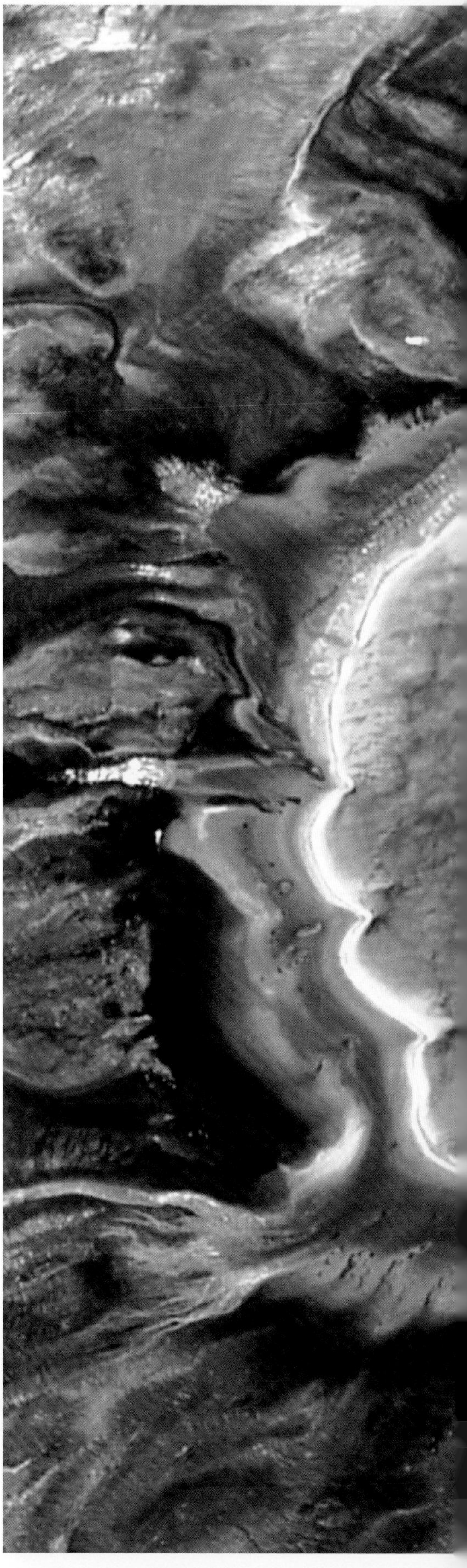

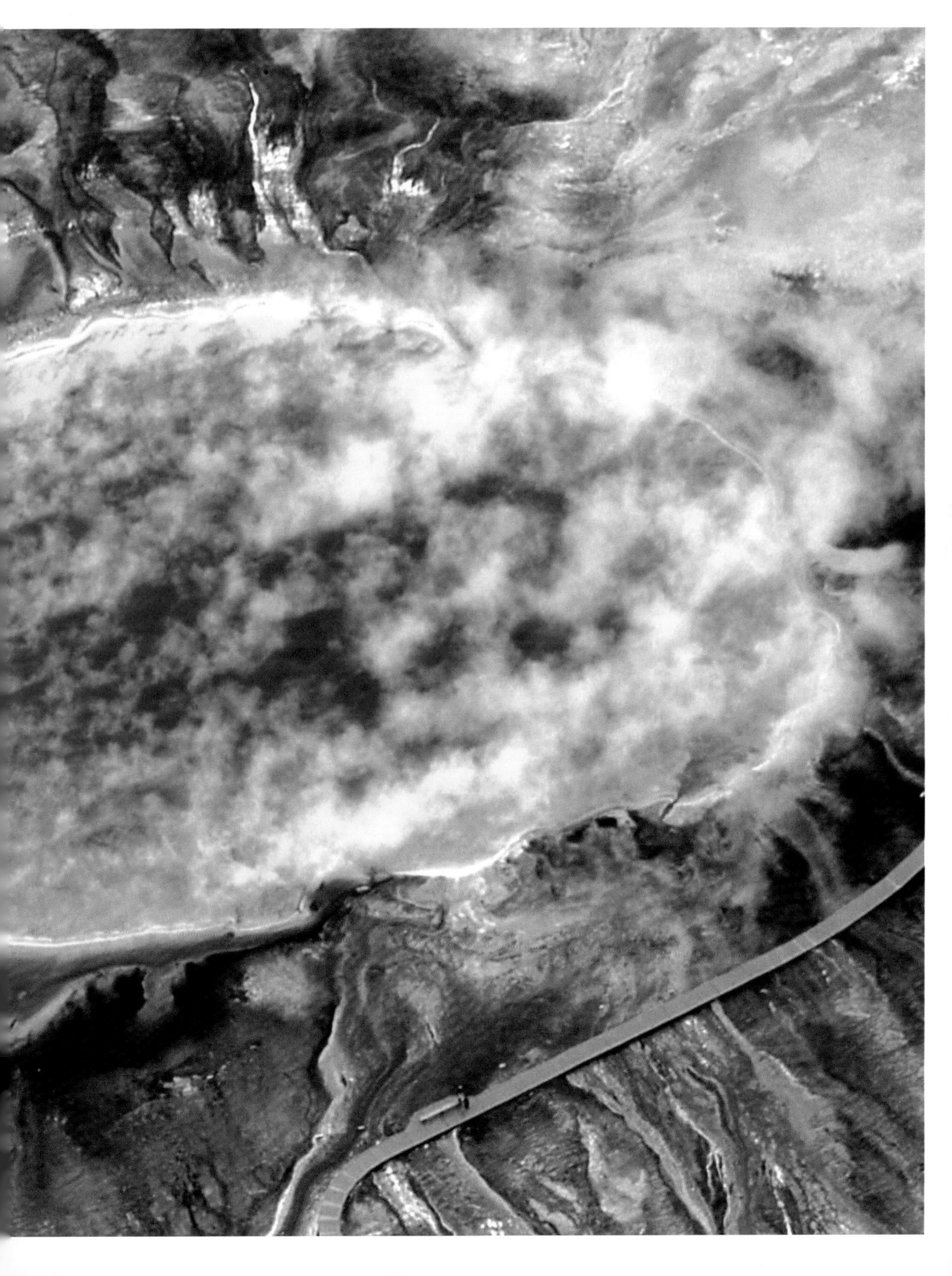

“

Dispatchers spoke with a man who had received an e-mail alerting him to a super-volcano eruption in West Yellowstone. He wanted to know if he should evacuate.
— Bozeman police reports, 2007